MY CELLS MADE ME DO IT

The Case for CELLULAR DETERMINISM

Why did I do that!? What was I thinking? Behavior is a curious phenomenon and we may often wonder why we do the things we do. In short, it's because of our cells.

Our behavior, the decisions we make and the actions we take, are nothing more than cellular responses. How our cells are interpreting their immediate environment dictates how we respond to our environment. Studies of behavior show that often we are unaware of why we make the decisions we do, and that these decisions can be easily manipulated and influenced by subtle environmental cues. What may seem like a decision made freely may indeed not be so. *My Cells Made Me Do It,* explains the phenomenon of behavior as a matter of cellular determinism.

"Why do people do what they do? This wonderful book explains the science and shows why it matters. We are our cells and this book helps explain the diversity of the human experience by starting with the cell."

—PAUL J. ZAK,
author, *The Moral Molecule*

"In the age-old debate over human free will, Robin Hayes presents an empirically based argument for deterministically fated behaviors. Insightfully drawing upon discoveries in molecular biology, his compelling case merges biology and philosophy to a new level in our pursuit to understand the human condition."

—ERIK GERGUS, PH.D.
Biology, Glendale Community College

"An answer to the age-old Problem of Freewill that demonstrates how and why our behaviors are inextricably determined by their biology and why free-will—or at least the illusion of free will—is necessary."

—TERRY D. JONES,
Professor of Anatomy, California State University, Stanislaus

Free-Will, Our Necessary Illusion

MY CELLS MADE ME DO IT

The Case for CELLULAR DETERMINISM

Robin R. Hayes

Moonshine Cove Publishing, LLC
Abbeville, South Carolina

SECOND EDITION
ISBN: 978–1–945181–98–6
PCN 2017956836

Illustration Credits
Shutterstock: 3, 4, 5, 6, 7, 8, 9, 10
Public domain Google images: 11, 12, 13, 14, 15, 16, 17, 18, 19

To my mother, Auberine Janell Nelson,
with love she put her fears into me.

ACKNOWLEDGMENTS

I'd like to start by thanking my wife Dawn and my daughter Meagan, to which both have contributed significantly to my experiences, for their love and support and their patience and tolerance as I obsessively ran on about Cellular Determinism. To my good pals Dr. Terry Jones and Dr. Erik Gergus for serving as sounding boards, allowing me to share ideas with them and for providing editing thoughts on those initial drafts. I'd like to thank my good friend Ken Whitaker for his thoughts on a very rough first draft. A special thanks to Mike Newton, his enthusiasm and excitement for the manuscript inspired me to pick it back up after I had let it sit idle for a time. I'd like to send a thank you to Elizabeth Tuttle for her comments and editing suggestions. I'd like to give an endearing thank you to Mary Goodrich for introducing me to epigenetics and for providing influence and inspiration along the way. A thank you to all the science and scientists that carryout the studies identified here and throughout. It is through their work that people like me can add our two cents to what it means. And a grand thank you to Gene D. Robinson and Moonshine Cove Publishing for giving my manuscript consideration. Finally, a thank you to all who have impacted my life: my family, my friends, my coworkers and my students. It is my experiences with all of you that have contributed to the writing of this book.

ABOUT THE AUTHOR

ROBIN HAYES is a microbiologist and Professor at Hartnell College in Salinas, California. With a graduate degree in Biology from Humboldt State University he has more than 25 years' experience in biological research, analytics, and education. As a research assistant at Stanford University's Hopkins Marine Station he studied bacteria associated with specialized squid glands and their eggs. He's developed education material for the Monterey Bay Aquarium, and has worked aboard a Korean processing vessel as a foreign fisheries observer for the National Oceanic and Atmosphere Association. Robin was the senior laboratory analyst for the Monterey Regional Water Pollution Control Agency during startup of the then largest water reclamation project for food crops in the nation. With the exception of a laboratory specific manual for his microbiology course, this is Robin's first publication. Forthcoming works include *Mostly Monogamous,* an examination of the social and evolutionary aspects of male-female relationships with regards to reproductive strategy.

CONTENTS

MY CELLS MADE ME DO IT

The Case for CELLULAR DETERMINISM

INTRODUCTION

> *"Things are the way they are, because they were the way they were."*
>
> —FRED HOYLE

This Is My Friend . . .

One night I was channel surfing when I paused for a moment on the situation comedy, *Mike and Molly*. Though having never seen the show, because of the commercials promoting it, I was familiar with the premise. Both Mike and Molly, heavy set people, were in a new and budding relationship. This particular scene caught my attention because of the subject matter unfolding. The scene has Mike holding a small plate of food while picking items from a snack table: Molly is standing to his left watching. From the stage set-up it is apparent they are attending an over-eater's or dieter's seminar. While Mike is completing his selections of snacks, a very pretty, full-sized woman approaches Mike and starts to flirt with him. Mike, flattered by the attention, clumsily flirts back. Just then we hear Molly clear her throat in an uh-umm "don't-forget-who-is-standing-next-to-you" kind of way. Mike, remembering

Molly is there, turns to the very pretty lady and says, "This is my friend, Molly."[1]

For guys, when a woman calls you a "friend" it can be the death nail in the proverbial coffin for any hope of a romantic relationship. The same can be true for women; calling them a friend, when they clearly think otherwise, can end any hopes of a romantic coupling. Being introduced as Mike's friend did not sit well with Molly. Mike finds himself in hot water and ends up trying to make amends for a mistake he may or may not completely understand. Mike's mistake, of course, is introducing Molly as a friend in an attempt to portray himself as available; just in case the pretty woman is interested. Mike's response seems almost automatic, instinctive, a reflex response. Clearly, little thought went into Mike's behavior. He did not think about the consequences until after the fact, even though, as revealed later in the show, he realizes he is lucky to be in a relationship with Molly.

The reason this scene caught my attention is it was not that much different from a scene in my own life some many years before. My wife and I met in college and had been dating during the fall semester. When the winter break rolled around we each went home for Christmas, but a couple weeks later she decided to visit me in my hometown. As I was showing her around town one of the places we stopped was a restaurant I had worked at during the summer. Shortly after walking in I realized I had to introduce this gal I was with, and, I wasn't sure what to call her. Of course, it's not that I didn't know her name; it's that people were going to expect some kind of title, a clarifier of the relationship. This restaurant, in particular, was an old stomping ground. If I introduce her as my girlfriend will I be hurting my chances with one of the other potential (unlikely, is probably more accurate) mates? I am not sure I was consciously thinking this, I just knew, even though like Mike I was in a pretty good relationship, that if I introduced her as my girlfriend it would imply I was off the market. Again, at the time, I do not believe I was consciously thinking this, only in retrospect. So, I did as Mike, and introduced my future wife as my "friend." As one might expect, this did not go over well! She did not say anything immediately, but by the time the new semester started I was single and we were just "friends." I spent the rest of the spring semester making up for this, as it turned out, not-so-minor faux pas.

I chuckled at the similarities of Mike's situation and mine so many years ago. But then I began to wonder, why would we respond so similarly? This must be a fairly common response if it is being used in a situation comedy. It is unlikely Mike and I had many common childhood experiences. Besides, our response seems more reactionary, without much thought. The simple answer then is it is in our genes. Years of evolution has programmed men to behave as pigs. Well, not really, but kind of; everyone understands that men are genetically programmed to spread our genes, to mate with as many different females as we can. At least that is what our biology instructs. Sperm is cheap to produce and once we start producing them, we do so for life. This is a very different strategy than females employ. Women have a limited number of eggs that are expensive to produce (relatively), and a limited number of years in which they are fertile.[2]

But what does it mean to be "in our genes"? When scientists talk about there being a gene for this trait, or a gene for that trait, they are rarely being literal. It is rare that a trait can be linked to a specific gene. Generally, traits and characteristics are influenced by many genes. But, again, how do genes influence traits and characteristics? Each gene is a sequence of DNA that codes for a specific protein. These proteins can be structural or enzymatic. They can turn on genes or turn off genes. They can cause reactions in the cell to occur, form surface receptors on the cell membrane, or communicate with other cells. Our genes are the language of protein synthesis. So, when men behave similarly, as dictated by millions of years of evolution, it's because their cells are producing the same proteins. These same proteins then operate on a similar level in most (we won't say all because there are always exceptions) men to produce a behavior that, at best, is frowned upon by women and, at worst, leaves one single.[3]

The proteins being produced by our genes operate at the cellular level. The response of the organism, in this case, men, begins at the cellular level. Across time, cultures, and experiences, the genes we share lead to similar cellular activity, which in turn leads to similar organismal behavior. The other commonality here, of course, is the situation: Mike's and my immediate perception of what was going on, i.e., pretty girl saying hi. The behavior of the organism is being determined by cellular responses to its immediate environment, both on the micro, and as we will see, on the macro scale. Could either of us have behaved differently? Given the situation, could we have made a different choice? Exercised our "free-will" to act counter to our biology? Were we already exercising our "free-will"?

Cellular Determinism

All of our thoughts and actions, moods and behaviors are the result of events that happen at the cellular level. These cellular events are determined by the environment of the cell and the cell's genetic makeup. No cell ever decides to respond to what is going on around it, it simply responds based on its programming (think billions of years of evolution). Therefore, our reactions to our surroundings are determined by the response of the trillions of cells we have, all responding in a determined manner to their individual environments. In fact, cells from near and far work to influence the environment of other cells in order to illicit a specific behavior.

Though we believe we are freely making choices, a careful study of biology will show we are only reacting as we can, as we must. A choice implies that we can do one thing or another with equal probability, or maybe there is a greater probability that we'll do one thing over another, but there is still a chance we could do the other. This is only an illusion. Choice is only an illusion. We react based on our cells programmed response to their environment. Our actions are determined and we take the only path we can.

This process of thought is not to suggest that the future is necessarily pre-determined; Quantum Mechanics and The Uncertainty Principle have shown there is enough randomness and uncertainty in the physical world to make the future unknowable. However, I'm arguing that our response to the physical world and the events going on around us are determined, but done so at the cellular level. What this also means is that every experience potentially influences the cells response, and therefore, our response.

We should also consider that anything that happens in the physical world that the organism does not perceive may have no influence on the cell and therefore no influence on the organism. Understand, of course, our senses are limited and environmental conditions may affect cells directly without the organism necessarily aware. That said, while the organism is responding to the physical world, it does not necessarily respond to everything going on. Still, everything the organism perceives influences the organism's path. Everything we see, hear, touch, feel, smell alters our nerve cells in some way causing a change at the cellular level that will forever influence our path. We call this, experience, and it is what gives us our illusion of choice.

When faced with making a decision we generally scan our memory and then decide on the best action. However, *we* don't really decide; our brain cells review all the incoming information and all the information on file (experience) and then the appropriate nerve cells fire. Our response to what is going on is happening at the cellular level and we have little control over those reactions. Our response is determined by what happens at the cellular level and what happens at the cellular level is influenced by experiences.

Some may find this revelation disturbing, that our paths are determined by our cellular responses to our environment, because it suggests we do not have free will. While it may be true that loss of choice seems to be the implication and that personal responsibility is diminished, this is not the case. Reality is subjective and so is illusion. Our free-will is both illusionary and real; a dualism we must acknowledge. Knowing that our every action, while determined by our experiences (and reading these words are now part of that experience) will have influence on all those around us, empowers us to be both freer with who we are and more active in our interactions with others. This information can help guide our behavior as we navigate the path put before us. Cellular Determinism is the biological process by which our behavior is derived. Recognizing this can aid us in understanding both our own behavior and the behavior of those around us.

PART ONE

RESPONDING TO THE ENVIRONMENT

CHAPTER 1

The Cell's Environment

"Timing has a lot to do with the outcome of a rain dance"

—COWBOY PROVERB

Crowding

In 2012, the San Francisco Giants won the World Series for the second time in three years. Even before my daughter was a year old I began taking her to games at old, windy Candlestick Park. Over the years, she and I attended many games together. Now, at seventeen, she was a big fan, very excited about them winning and wanted to go to the championship parade being held in San Francisco. So, blowing off class for the day, the two off us jumped on a very crowded train headed to the city.

On this day, San Francisco was being inundated by thousands and thousands of adoring fans. While we were able to move around easily enough initially, as the crowd grew our movements became more and more restricted. Soon, all we were able to see and hear was being dictated by those around us. If you have ever lived in a big city, been to a sporting event, or attended a concert, you have likely experienced a large crowd. When crowds get really large, the individuals within those crowds become restricted by the crowd. That day in San Francisco started with us being able to walk around looking for a good spot from which to watch the parade, finding one, and settling in. As the day wore on and the crowd increased in size we found our view of the parade route being narrowed; we found our personal space being breached as people crammed together; we found ourselves being exposed to odors that other individuals were producing. Much of what we were able to see, our movement, and the smells we were experiencing, were the result of those around us. Our experiences were being dictated by those in our environment and whether we realized it or not, so were our responses.

It's All About Timing

Bacteria, when grown in a laboratory, behave in a very predictable manner. They start out, upon first being introduced to a new medium, such as food source or anything bacteria use for energy, just hanging out and synthesizing proteins. You can think of this as a getting-used-to-your-new-neighborhood period. The microbes don't start growing immediately, they assess the available food source and start making the enzymes that will be necessary to break down and use this fuel. When we say growing, in reference to bacteria, we're not necessarily talking about increase in size, although that does happen, we are mainly talking about an increase in number, a multiplication of cells. Not all bacteria can use all fuel sources and so the type of medium will dictate the kind of bacteria that will grow. That being said, any bacterium being transferred to a new, suitable medium will experience this lag phase in growth.[4]

After some time has passed, the bacteria will begin to divide. How much time? That will depend on the growth conditions established and the microbe itself, but divide they will. Initially, bacteria, while not growing, are actually very busy making energy and macromolecules that will be used for the building of news cells. Once the lag period has passed, bacterial growth sky rockets! This phase of the bacterial growth curve sees the microbes growing exponentially. Their doubling time, a measure of how fast the cells are dividing, is at its fastest pace.

Known as the logarithmic phase, the bacteria are utilizing the available food stuff at a rapid pace in their efforts to divide. Of course, this level of growth can only continue if the energy source is unlimited. In a test tube or on a petri dish of solid medium, the nutrients are limited and so this phase of growth will come to an end as the nutrients become depleted.

Soon, bacterial growth slows; fewer and fewer new cells are being born and those alive are now dying at a steady rate. You see, in addition to there being less to eat, bacteria in a closed system like this also have to deal with their own waste products piling up. They have no garbage man to carry away end products, many of which can be toxic to the microbes as the levels increase. Before long the number of viable cells is greatly reduced. Eventually the culture will die. No viable cells will remain.

This growth pattern, or growth curve (Figure 1), is essentially the same for all bacteria in a laboratory setting. If growth conditions are right, i.e. medium, temperature, etc., all bacteria, after being transferred

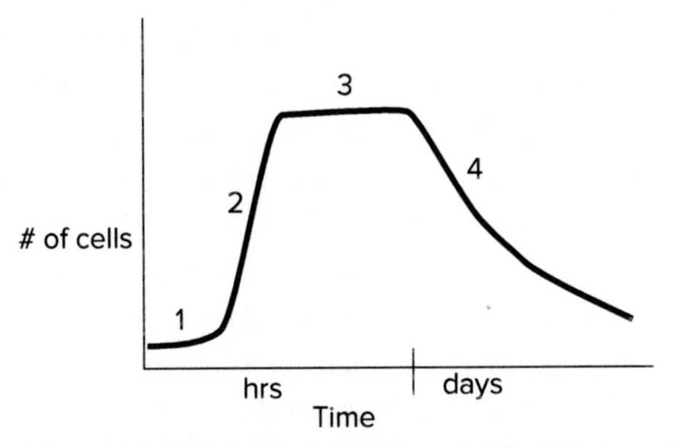

FIGURE 1: Bacterial Growth Curve: 1. Lag phase; acclimating to environment. 2. Log phase, conditions great for growth. 3. Stationary phase, environment lacking nutrients and accumulating waste. 4. Death phase, microbes can't survive conditions.

to a new medium, experience a lag phase followed by a log phase followed by a stationary phase followed, finally, by a death phase. The pattern is highly ordered and predictable.

When bacteria multiply, they do so by a process known as binary fission. They simply increase in size, replicate their genetic material and then split in half. The two resulting daughter cells are genetically identical. In fact, all the cells that make up an isolated colony, growing on an agar plate or in a test tube, are genetically identical. They are all descendants of the original cell by this process of binary fission, an asexual form of reproduction. They all have the same DNA, they code for the same enzymes and generally will behave similarly. That is, depending on environmental conditions, they will utilize the same nutrients and produce the same products.

Now, of course, mutations occur and it is quite possible, even likely, that the specific nucleotide sequences for any two cells might not be exactly the same. However, mutation, as a result of nucleotides being substituted one for another, resulting in new protein products, is a rare occasion. We will see later where such an event, new protein products from spontaneous mutation, has been observed, but for the most part, it is safe to assume all cells of a colony are identical with regards to their enzyme and protein capabilities.

Imagine now, a bacterial cell being "born" into a colony on a solid agar medium. The type of environment will vary significantly for each cell of the colony depending on when one is born. Remembering that a colony consists of millions of cells, all derived from a single cell and all genetically identical, the eighth cell of that colony will experience very different conditions than the one millionth cell. The eighth cell being born is entering when conditions are optimal for growth. There are lots of nutrients, plenty of space and not much waste. Also, remembering our bacterial growth curve, these early cells will start feeding on the available nutrients and begin dividing. Conditions being ideal, these early cells may divide many times.

On the other hand, if you are the one millionth cell being born, conditions might be slightly less than wonderful, depending on where in the colony you are born. If you're on the edge of the colony nutrients may still be abundant as you move away from the colony. Additionally, waste products will be minimal here on the edge. Cells on the edge may continue to divide as the colony grows. However, their number of division may be substantially less than that of the early cells.

If you are the millionth cell being born somewhere in the middle of the colony, conditions could be horrible. You may find yourself where nutrients have been depleted and there is a high concentration of waste products. Some bacteria produce acid as waste, ultimately, completely changing the pH of their environment. A cell being born here may not be able to survive, let alone divide.

Therefore, although the cells of a colony are all equipped with the same raw materials, the cells themselves are going to respond very differently based upon the environment they are being born into. The early cells go straight to growing and dividing. Cells on the edge put less energy into dividing and more into moving to better pastures. Those in the interior of the colony are likely to not divide at all and may actually start forming resting stages such as endospores or cysts.

Why the different responses from these genetically identical cells? Well, obviously because the environments are different, specifically, the molecular environment immediately surrounding the cell. We call this the microenvironment and it is of great significance to the microbe. It is these immediate molecules that influence how the cell is going to respond. This molecular world around the cell will determine what the cell does. In good times the cell will divide; in areas of depleting nutrients and increasing waste the cell may become mobile; and, in places of no nutrients and high waste, the cell may sporulate or die.

How does the cell know what to do? How does it choose to divide versus not divide? Each action requires the synthesis of very specific enzymes, enzymes that are coded for by the genes all these cells share. The response from the cell has everything to do with the types of proteins found on the surface of the cells. If these cells are genetically identical, shouldn't they all have the same surface proteins? Initially, they may have, but these proteins will interact with the molecules of their surroundings and, because molecules are not necessarily evenly distributed, it is likely more of one type of receptor, over another, may become bound. For instance, a cell may have equal numbers of receptors for glucose and lactose initially, but if no lactose is present in the environment than those receptors do not become attached to anything. So the cell will be responding to both the presence of glucose, because receptors are being bound, and the absence of lactose, because receptors are not being bound. These interactions will determine what types of enzymes the cell will produce. The environment is dictating which genes are active and which genes are not.

Let's pause for a moment and consider the make-up of a bacterial cell. The outer structure, the cell wall, is composed of molecules that are uniquely bacterial. This structure is somewhat rigid (providing the cell its shape and preventing the cell from bursting) and permeable (allowing water and other molecules to pass freely). There are two fundamentally different cell wall types, each with their own uniqueness regarding molecular structure and interactions with the surroundings.

Just inside the cell wall is the cell membrane. This structure is very similar in most cells. The cell membrane is a two layered structure formed from molecules called phospholipids. These molecules have water-loving heads and water-hating tails and are arranged such that the water-hating tails face each other. This creates a very fluid, dynamic structure that serves to allow water to pass freely, but regulates the movement of other molecules across this barrier. Embedded within the cell membrane are lots of different proteins and enzymes with various functions. Some proteins serve as channels to allow molecules to pass in and out of the cell, facilitating the transport of nutrients in and waste out. Many other proteins combine with sugar to form glycoproteins that serve as receptors.

Receptors for what, though? Well, receptors for the molecules in their immediate environment. How does the cell know what molecules will be in its immediate environment? It doesn't. Its genes code for specific types of receptors that will be specific for certain molecules. If those molecules are present they will bind to these glycoproteins, which through a series of chemical reactions will relay information to the cell's DNA. This information, in turn, will determine which genes may be active and, subsequently, what type of proteins might be on the surface.

The cell membrane surrounds the gel-like cytoplasm containing the genetic material of the cell in addition to the machinery necessary for the surviving, growing, and dividing. The genetic material of bacteria comes as a single circular piece of DNA. In addition to this single chromosome, many bacteria have extra circular fragments called plasmids. Plasmids will often impart special capabilities to a cell it might not otherwise be able to do. For instance, plasmids may code for an enzyme allowing the cell to destroy certain antibiotics that those without the plasmid would be susceptible to. Sometimes plasmids can lead to toxin production.

The compilation of the cell's single chromosome and any plasmids it might have are responsible for the types of receptors found on the cell's surface and the enzymes it produces. So, it is this genetic material providing the proteins to which the environment will interact, and this interaction will lead the cell to respond by producing and releasing certain proteins. Some of these proteins may be increases in specific types of surface receptors, allowing the cell to better respond to environmental changes; others will be enzymes likely designed to take advantage of the present environment.

Let's return to the cells of our colony. If you are one of the early cells, say the eighth, you are coming from a mother cell that has been living in good conditions. Since you will be getting half of her cell membrane when she divides, the surface receptors you start with will be the ones she was synthesizing at the time. If conditions are good, you probably have many receptors or protein channels for the type of food stuff available. You are also probably producing the enzymes necessary to breakdown this available nutrient. You still have the genes that code for enzymes and surface receptors to take advantage of different food stuffs, but those genes are not active and the surface receptors are fewer.

Suppose, for a moment, you were lifted, shortly after being born, from the medium your mother had been growing on to a medium very different. Maybe your mother was growing on a medium rich in glucose and now you are on a medium rich in lactose. None of your glucose receiving surface receptors are being bound. Instead, the few lactose receiving surface receptors you have are now being bound by the lactose in the media. This environmental change is going to be relayed to your genetic material and you are going to respond by synthesizing more lactose receiving surface receptors and more lactose metabolizing enzymes. Of course, if you do not have the genes for lactose utilization then you likely die.

Let's get back to being the eighth cell on some nutrient rich medium. Since your mother was already adjusted to these great conditions you were born with the surface receptors and enzymes ready to take advantage of such conditions. You may start dividing very shortly after being born because you are already equipped to do so and the environment is right. In fact, all the cells of these early generations are likely dividing and at their greatest rate. This is why we see the exponential growth described in the bacterial growth curve.

If you are the millionth cell being born, be it on the edge of the colony or in the middle, environmental conditions will be quite different than what the very early cells encountered. Additionally, your surface receptors and enzyme production are also different than what the very early cells might have had. The concentration of certain receptor types may have changed; you may have more receptors for one type of molecule (lactose) and less for another (glucose). You may have new receptors, coding for specific molecules that the earlier cells might not have encountered. The presence of specific surface receptors allows the microbe to detect where the food is and where the waste is coming from. Known as chemotaxis, this allows your movement to be somewhat directional. So, while still wanting to divide the cell has to spend a little more energy getting food, and thus the number of divisions will be far fewer than the eighth cell.

Each of the cells described above share the same genetic material but each responds in very different ways. Their behavior, if you will, is determined by the interaction between the environmental conditions and the genetic material. None of these cells choose to divide or move or form endospores, they do so because the environment dictates it. Of course the environment has to work in concert with the genetic material; if you don't have genes that code for the surface receptors to bind to environmental molecules then those molecules do not influence your response. That said, as a cell, your only possible response is the one that is determined by your microenvironment. No decisions are being made; you are just reacting, responding to what is going on all around you. You only have one path.

This type of determinism, not entirely environmental and not entirely genetic, is what guides all cells in their behavior. I call it Cellular Determinism; it applies to not just bacterial cells of a colony growing on an agar medium in a laboratory, but to all bacterial cells, worldwide. Bacteria of the ocean, bacteria inhabiting soil, bacteria living within our intestines, are all assessing their immediate environment using surface receptors to respond to existing conditions. These factors do not apply to just bacterial cells (prokaryotes) but also to eukaryotic cells—those with a nucleus. Although more complicated than bacteria, single-celled organisms like algae and protozoa also have surface receptors they use to respond to the environment. Certainly one wouldn't suggest that these simple, single-celled organisms are considering options with potential outcomes when they swim or feed or divide. No, these single-celled eukaryotes are responding as determined by their environment and the current surface receptors they possess.

However, it doesn't stop there; Cellular Determinism applies not only to single-celled organism like bacteria and protozoa, but also to multicellular organisms. The cells that make up any organism; fungi, fish, frogs, ferns and Francines are responding to cues from their microenvironment. Again, one wouldn't suggest that an individual cell of a multicellular organism is giving any thought to how it should respond to existing conditions. All each cell does is assess the environment based on the molecules present and then respond accordingly.

Evolution has led to cells of multicellular organisms responding in a coordinated fashion such that one would assume choice is being made. A frog chooses to jump, a dog chooses to chase a ball, and a woman chooses a mate. But, at the cellular level, no choice is being made, each cell is responding to the molecules surrounding it. The cell's path is determined. The path for all the billions or trillions of cells an organism may have is determined. The path of the organism is determined by the coordinated response of these trillions of cells. The organism does not choose how it will respond, it simply responds.

CHAPTER 2

The Autonomous Cell

"Any living cell carries with it the experience of a billion years of experimentation by its ancestors."

—MAX LUDWIG HENNING DELBRÜCK (1949)

Individuals in a Crowd

Going back to parade day in San Francisco, my daughter and I found ourselves, in many ways, at the mercy of the crowd. Many times, if there was a push or surge in the crowd, we would be moved with it. Bumping into people and rubbing against them was common and became even more so as the crowd grew. Various aromas from the activity of individuals around us filled our olfactory sense. Much of what was happening to us had everything to do with the fact that we were part of this much larger group.

Still, we were individuals among this crowd. While it was true that what we were experiencing was being influenced by those around us, we were still able to somewhat manipulate our experience by our reactions. It took some considerable effort, but we could push our way through the crowd to more open spaces. We could stand on our toes or lean forward to get a better view of some spectacle. And, if the odor was too offensive, we could hold our breath or nose until it passed. Yet, even these behaviors were being dictated by the crowd. A very similar situation exists for the cells of a multicellular organism. While the cell's activities are greatly influenced by the environment being created by neighboring cells, each cell is still functioning individually—growing, metabolizing, synthesizing.

The Cellular Entity

It's amazing to realize that the human body is composed of trillions of cells, all genetically identical and all derived from a single cell—not unlike the cells of a bacterial colony. It's just as amazing to realize these cells are each functioning in a specific and coordinated fashion to cause the organism to respond to its surroundings. While working in concert, each of these cells is still very much its own entity, responding to the immediate environment based on the types of surface receptors each possesses. If we manipulate the environment, we can manipulate cell behavior. Change the salt content, adjust the pH, add various chemicals to the environment and cells will respond to these changes. We are essentially a collection of trillions of cells, each reacting independently to their own microenvironment. Cells can influence each other based on chemical releases and physical contact—that's how we respond in a coordinated manner—but each cell's response is based on its own receptors and the molecules surrounding it. Our behavior is dependent on a collection of fully independent cells.

If you doubt each cell is its own entity, consider for a moment the cells of Henrietta Lacks. You may be familiar with the story. In 1951, Henrietta Lacks was diagnosed with cervical cancer. During her treatment some of the cancerous cells were removed and sent to the lab for study. Surprisingly, the cells lived much longer than the normal few days most cells were able to survive at the time. In fact, these cells were easy to culture and grew well. Soon, cells from the HeLa (short for Henrietta Lacks) line were being sent to labs around the globe for research. For the next 60 years HeLa cells were used in labs all over the world for scientific study in areas such as polio, AIDS, gene mapping, and many more.[5] These cells, long separated from the organism, many, likely never part of the organism, are still growing strong today. The cell line that was established when Henrietta's parent's sex cells fused together (egg and sperm combining) has essentially become immortal. Does the individual cell really need the organism?

What causes a cell to become cancerous and why are they immortal? All cells have genes that control when that cell will divide. Cancer cells have essentially lost the ability to control cell division. As you might expect, these genes are influenced by surface receptors and what those receptors are bound to. Under normal conditions, contact with neighboring cells can be enough to inhibit a cell's division. In cancer cells the genes that control when a cell will divide have mutated and cell division becomes unregulated. Along with this uncontrolled division generally comes a loss of function. As a result, there grows a clump of non-functional cells—a tumor.

Cells that divide frequently, epithelial cells like skins cells and lung cells have a greater chance for mutations to occur during the replication of their DNA. The more times a cell replicates the more chances mistakes will happen. When a mutation to the oppressor gene, a gene that turns off cell division, occurs, it can cause the cell to lose what is called contact inhibition. Cell division is no longer influenced by the presence of other cells and uncontrolled division ensues. Mutations like this can occur randomly during the replication process, which is why cells that divide more often have a greater chance for developing cancer.

These cells can also be environmentally influenced. In 1984, Harald zur Hausen, discovered that the Human Papilloma Virus (HPV) was the major agent for cervical cancer.[6] The presence of the virus can cause a disruption in the cell's replication process leading to the mutations that give rise to cancer. It is not surprising to learn that Henrietta Lacks was infected by HPV–18, a particularly harmful strain.[7]

Why are Henrietta's cells, and cancer cells in general, immortal? For most cells, at least those of the human body, it is suggested they undergo 50 divisions,[8] more or less, during the life of the cell. Once the cell nears this number, it stops dividing. One of the reasons we grow old is our tissues stop renewing themselves via cell division.

What causes a cell to lose the ability to divide? It turns out that at the end of our chromosomes is a string of nucleotides called telomeres. These segments of DNA are analogous to the little plastic coverings at the end of shoestrings; they serve to protect the end of chromosomes from degradation during replication. However, replication of the nucleotide sequence cannot go to the end of the DNA molecule, so the telomere nucleotides on the ends are lost. When a cell divides, its string of telomeres gets shorter. When all of the telomeres are gone, the chromosome can no longer be replicated without loss to the ends. This impairs those genes on the end and leads to the end of cell division and eventually cell death.[9]

With cancer cells we don't see this same shortening of the string of telomeres. Most organisms are capable of producing the enzyme telomerase that can add telomeres back on to the end of chromosomes, but normally do not do so; however, cancer cells do.[10] So, not only are they dividing uncontrollably, but they are also circumventing the programmed death built into the system. Imagine if we could turn on and off telomerase activity. Could we live forever?

What this indicates is that such immortality is possible for each and every one of our cells. All of our cells have the same DNA with the same potential to become cancerous and the same potential to be immortal; it's just a matter of turning on the right switches. Henrietta Lacks' cells have been alive long after the organism has died. Will these cells live forever? Who's to say? Theoretically, it's possible. What we do know is they are going strong 60 years after Henrietta's death.

All of our cells respond to their external environment, being influenced by the molecules of other cells of the organism, they carry out their specific function to the benefit of the body. When the machinery malfunctions, each cell has the potential to go rogue, behaving selfishly to the detriment of the body.

Cell Conversion

As one more piece of evidence to the potential of any cell, consider the recent advancement in cell conversion. It has been a long and widely held idea that once cells differentiated into whatever they were to become, i.e., skin cells, nerve cells, fat cells, etc., they could not become something else. There are, however, some cells, called stem cells, with the ability to give rise to other types of cells. Embryonic stem cells have the most plasticity and can give rise to most other types of cells, but all stem cells have limits in what they can give rise to. Basically, at some point in development, a cell's fate becomes determined and fixed. Even though they share the same DNA, skin cells do not become nerve cells, except when they do.

In 2007, Dr. Kazutoshi Takahashi and his team were able to reprogram dermal fibroblast cells into induced pluripotent stem (iPS) cells.[11] A pluripotent stem cell is one that can give rise to many types of other cells. By manipulating the environment of the cell, Takahashi's team was able to convert the cell from its programmed fate to a cell with the same potential as embryonic stem cells. The researchers used what are known as transcription factors—proteins—to turn on or off certain genes. These induced pluripotent stem cells expressed similar telomerase activity and cell surface markers as embryonic stem cells. It shouldn't be surprising that the cell's surface markers would change to better resemble embryonic cells; if indeed conversion has taken place. Upon conversion, the cells must change from having receptors used by skin cells to do skin cell stuff, to cells having receptors that must read the environment in preparation to become whatever cell type the environment might dictate.

The idea with induced pluripotent stem cells is that once you convert a cell to this pluripotent state, you can now manipulate the cell to become a liver cell, or a heart cell or even a brain cell. Flood these pluripotent cells with the right chemical bath and you can induce differentiation. In this way we can potentially repair tissue damage by inducing growth of iPS cells.

In 2009, Dr. Thomas Vierbuchen took the technology a step further. Using the same concept as Takahasi, Vierbuchen wondered if transcription factors might be able to convert cells directly; that is, bypass the creation of pluripotent cells. After weeding through nineteen possible transcription factors, Vierbuchen and his team hit on a cocktail of just three factors that had the ability to change fibroblast cells directly into nerve cells. The combination of just three transcription factors was enough to convert a cell functioning to secrete intracellular connective tissue to one able to produce proteins specific to nerve cells, conduct electrical impulses, and form synaptic connections with other cells, becoming functionally a nerve cell.[12, 13]

What are these transcription factors (TF) and how can they convert cells? The TF molecule is a protein produced by the cell, or maybe neighboring cells, that binds to specific regions of the DNA molecule. Once bound, these proteins serve to regulate whether a particular gene will be on or off. This is done by controlling RNA polymerase. These molecules bind to sequences on DNA and transcribe the genes into RNA molecules. These RNA molecules will then be translated into functioning proteins, like transcription factors. Transcription factors control the binding ability of RNA polymerase to the DNA molecule either by blocking the binding sequence the polymerase molecules would bind to, or by promoting the recruitment of the polymerase molecule. So, the presence of specific transcription factors will determine whether any particular protein is being produced, including the production of other transcription factors. This means the presence of any one transcription factor can lead to the turning on or off of a number of genes. This, in turn, leads to a change in surface receptors and ultimately a different response to the environment.

This technology is in its infancy. At the time of this writing a new study reports the conversion of umbilical cord stem cells into brain support cells;[14] another study reports changing fully formed liver cells into functional nerve cells.[15] The significance of these studies is they demonstrate conversion can occur across embryonic germ layers.[16] Theoretically then, any cell has the potential to become any other type of cell. All one has to do is hit on the right combination of transcription factors—turn on or off the right combination of genes—and you should be able to convert any cell into another. Of course, the tricky part is finding the right chemical environment.

Clearly, while each cell of a multicellular organism is acting together with the other cells of that organism, their responses are still made as individuals. The cell is responding independently to the molecules found both in and around it. It is the independent actions of these trillions of cells that will determine the behavior of the organism.

CHAPTER 3

Cellular Influence

"The cell never acts; it reacts."

—ERNEST HAECKEL (1866)

Mary and Steve

Given the independent nature of cells, how might their responses to environmental cues influence our behavior? Consider for a moment, Mary and Steve. They married three years ago and have been together exclusively for the past five. Mary met Steve during her junior year in college. He was kind and good looking and just made her feel comfortable. Mary had been sexually active almost as soon as she got to college and had been taking birth control pills since her senior year in high school. She liked being on birth control; her periods were regular and she felt somewhat liberated, sexually anyway. Mary had slept with several guys before she met Steve, but once she fell in love with Steve, nobody else even caught her eye.

Recently, Mary and Steve have decided to start their family. Steve has a good job now and job would allow her to take maternity leave.

They have their own house and the timing seems right. So, Mary has gone off birth control.

Although she hasn't made the connection herself, since going off birth control Mary has had an ever increasing attraction to Sam, a co-worker in her office. There is nothing particularly striking about Sam; he is good looking enough, but so is her husband. However, she finds herself shamelessly flirting with Sam. When she stands next to him at the copy machine, or when they get coffee together, she has an overwhelming attraction to him. She finds herself flush and actually contemplating having a drink with Sam. She would never dream of cheating on Steve, except, she is dreaming of cheating on Steve. Mary can't understand this because she loves Steve very much, but the desire to have sex with Sam is quite intense.[17]

MHC Molecules

What Mary doesn't know is that her body is telling her who her better mate is, molecularly anyway. Even before the microenvironment of the egg and sperm are determining which sperm will do the fertilizing, the microenvironment of the cells of the individuals are influencing which two people will get together..

Located on the surface of our cells is a grouping of protein molecules called the major histocompatibility complex (MHC). This molecule is of extreme importance to the body's immune response and overall self-recognition. All of our cells—with the exception of red blood cells—possess MHC molecules. These molecules function to present antigens to our white blood cells, which in turn interpret the antigen and determine the next course of action. There are two classes of MHC molecules, but except for some specific white blood cells, all of our body cells have MHC class 1 molecules.[18]

These MHC class 1 molecules function to present internal molecules to our immune cells. It is a way of monitoring the health of the cell. The cell will sample molecules from within and place these in the MHC molecule; much like a hotdog in a bun. Then the white blood cells known as cytotoxic T-cells will scan the surface of the cell, reading the molecules being presented. If the molecules are recognized as self, or not seen as a threat, then the cytotoxic T-cell moves on to the next cell.

However, occasionally a cell will become infected by a virus or bacterium, may become cancerous, or have some other abnormality. In presenting these endogenous (internal) antigens, there is a good chance that viral proteins or abnormal proteins will be presented. If a cytotoxic T-cell, while scanning the MHC molecules of a cell, comes across a protein it does not recognize or interprets as dangerous, it will release chemicals that will cause the abnormal cell to destroy itself. The cytotoxic T-cell is using its interaction, this time physical contact, to determine if a cell is healthy and if not it is then using chemical messengers to tell the cell to die. The body uses this system to control the spread of disease from infected cells to uninfected cells and to kill cells that may have become cancerous.[19]

In addition to the above role of antigen presentation, MHC molecules also play a major role in self recognition. Each T-cell, be it cytotoxic T-cell or helper T-cell,[20] is first tested by the body to see if it can recognize MHC molecules. The two major cells of our immune system rely on these molecules to determine what is self and what is not self. Those T-cells that do not recognize self MHC molecules will be destroyed. Again, this is an extremely significant function being performed by the simple presence of specific surface proteins. MHC molecules allow our immune system to recognize both the cells that are fine and the cells that are not so fine.

You Smell Great

So, why am I telling you about MHC molecules? As it happens, these molecules, protein components of all cells, seem to play a role in a woman's choice of mate. Several studies have shown that women have a preference for men depending on their MHC dissimilarity. A study conducted by Claus Wedekind,[21] demonstrated that women seem to prefer men who are less similar with regards to their MHC molecules. Wedekind typed the participants of his study based on their MHC molecules. He then had six males of various MHC types wear t-shirts for two days. The following day he had the women smell the t-shirts and rank them in order of pleasantness. He found the women were more attracted to the odor of males with MHC molecules different from their own then they were to men with MHC molecules similar to their own.

The story doesn't end with these MHC molecules. In addition to finding the odor of men with dissimilar MHC molecules more pleasant, the test subjects stated that these odors reminded them of men they were either currently dating or had been with in the past. The women here are detecting specific molecules using their olfactory sense, which in turn, triggers a behavioral response; attracted or not attracted—potential mate, or not so much.

It also seems that oral contraceptive can have the effect of altering the MHC molecules a woman prefers. In 2008, Craig Roberts and team published a study suggesting that a woman's preference for MHC molecules can change when she is on birth control pills.[22] The men with MHC molecules more similar are now 'smelled' as the more desirable.

What does the pill do to alter a woman's preference? Birth control pills release hormones that make a woman's body act as though she is pregnant, preventing ovulation. Various levels of estrogen and progesterone will interplay with the body's cellular surface receptors, causing those cells to act accordingly, in this case, mimicking pregnancy.

It has been suggested the reason women may prefer men with dissimilar MHC molecules when not pregnant may be twofold. Firstly, choosing someone with dissimilar MHC molecules to mate with can have the effect of increasing heterozygosity (the scientific way to say it increases genetic variability). This increased variability can have the effect of improving the immune response and keeping the population one step ahead of pathogens. Secondly, preferring dissimilar MHC molecules can serve to limit or prevent inbreeding;[23] a practice that can cause deleterious genes to be manifested.

Right Guy Wrong Time

So why does her preference change when she is pregnant, or 'perceived' pregnant? The suggestion is that men with similar MHC molecules may relay a feeling of familial familiarity. If she is pregnant, she is no longer looking for a mate, but now looking for stability. Those with similar MHC molecules are now 'smelled' as the more desirable because they are more likely going to aid in the raising of the child. She associates their smell with family, which is generally a feeling of comfort. A woman's preference for men changes depending on her receptivity to mating.

One can see the evolutionary advantages gained by mating outside her pair. The woman gets the more dissimilar genes, increasing variability and decreasing inbreeding, while also gaining from the stability of familial relationships. Many studies have shown the propensity of females, of numerous species, to step out, if you will, on their long term mate for the potential of better, or in this case, more dissimilar, genes.[24]

While this recognition process for MHC molecules has been advantageous to species in the past, recent modern developments have thrown a wrench into the works. Today, many women all over the world are on some kind of oral contraceptive, thus they are going around in a physiological state of pregnancy. In this state, women may find men with similar MHC molecules to be more desirable. That desirability may not be sexual, per se, remember, physiologically she is pregnant; her attraction to these men is because they feel safer—and when one is pregnant, safe is preferable.

Imagine now, a woman becoming involved with a man she finds desirable because she is on the pill and prefers his more similar MHC molecules. They decide to have a child and she goes off the pill. Soon, because her body is no longer behaving physiologically as if it's pregnant, her preferences, for mating anyway, switch to guys with MHC molecules that are more dissimilar. Now, for the first time in her relationship, she finds herself attracted to other guys. Her mate of choice is no longer her choice mate.

A 2006 study by Christine Garver-Apgar et al.,[25] found that women sharing MHC molecules more similar to their partner's, were more likely to cheat. The study looked at 48 couples, together at least two years, who described themselves as exclusive (they were either married, living together, or dating only their partner). These couples were asked questions about their relationships, their sexual feelings towards their partner, and their faithfulness. While the study didn't look at the number of women that may have begun dating their mate while on the pill, it did determine the relative similarity of MHC molecules shared between mates. Comparing the answers on the surveys with the relative similarity of MHC molecules shared between couples revealed what we might have predicted. As described by Dr. Garver-Apgar, as the similarity in MHC molecules increased between the couples, the women "were less sexually responsive to their partners, more likely to have affairs, and more attracted to other males, particularly during fertile days of their menstrual cycles"[26] (Figure 2).

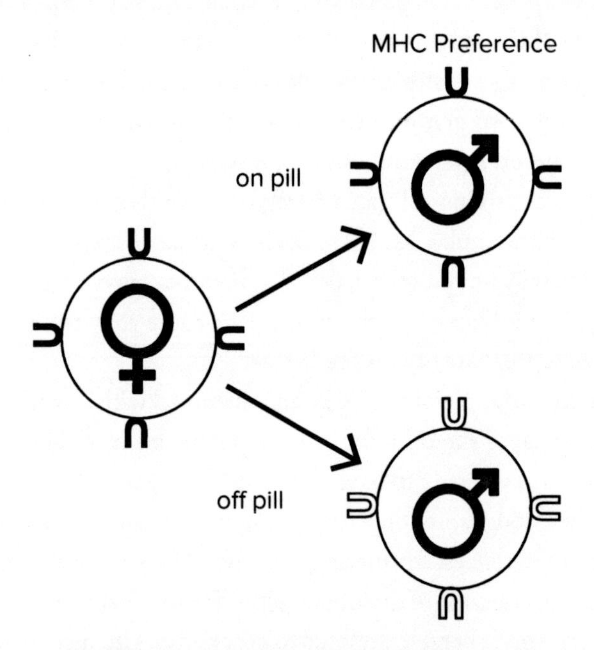

FIGURE 2: Illustrating how her MHC preferences change depending on her physiological state.

"When MHC genetic differences were significant between partners, cheating behaviors were either absent or greatly reduced. MHC sharing and the woman's number of adulterous partners were directly linked. When the partners had 50% of the MHC genes in common, the women had a 50% chance of cheating on them."[27]

So, what are the implications here regarding the interactions at the cellular level, between cells and the molecules of their microenvironment that determines the actions of the organism? In general, in today's society, stepping out on your spouse is viewed negatively. Faithfulness in a relationship is expected. But for some women, in some relationships, is this really possible? Nature—in the case of cellular biology—is a powerful force. If a woman's cellular biology is driving her to find a suitable mate, and she is not currently with that mate, can she realistically resist? Many will say, "Of course she can! Our consciousness, ability to think, free-will, gives us the capability to override our natural urges and impulses." But do they, considering they too are all part of the same process? She is responding to her cellular biology, her actions are determined by her cell's surroundings. She will only be able to resist if there is a greater, overriding, cellular influence.

Of course there are many factors that influence mate choice in addition to smell. Most people would rank visual attractiveness high. A recent study seems to contradict the relationship between MHC molecules and mate preference. In 2005, S. Craig Roberts et al.,[28] released a report showing women preferred the faces of men with more similar MHC molecules. The women of the study were shown pictures of men and then rated them based on attractiveness. Those men with more similar MHC molecules to the particular women rating them were viewed as more attractive.

While on the surface this seems contradictory, it is not unexpected. Again, mate choice is governed by many factors, visual pleasantness usually being one of the first and more important cues. Liking our own face and growing up around people who look like us, we associate certain facial features as comforting, stable, attractive. So, of course we might find these features more appealing. This establishes a pool of mating candidates from which to 'choose' based on visual preference, but as we will see later, emotional decisions like sexual activity require input from our olfactory sense as well. Finding a face attractive can happen because of familiarity or relation, but to insure variability, olfactory preference for dissimilar MHC molecules comes into play.

Mixed Signals

The pill was introduced to the American public in 1960 and within five years 6.5 million women were taking the oral contraceptive. Despite a brief, temporary decline in the seventies due to questions raised about possible side effects, usage continued to grow until by 1980 nearly 11 million women were using the pill.[29] At the same time, America's divorce rate began to increase. Starting in 1965, after having been fairly stable for years, the divorce rate for women over 15 years of age more than doubled by 1979.[30] Correlation is not causation and there can be a number of socio-economic reasons for divorce, however, is it that much of a reach to suggest the increase in divorce could be the result of women, with confused physiology, choosing the wrong men? Can it be (based in part on MHC molecules and their influence on a woman's preference) that the pill has indirectly (or directly) led to our increased divorce rate?[31] A woman's body thinks she is pregnant while she is on the pill, her body chemistry is altered, and this altered state leads her to make a decision she might not have otherwise made.

The point of my digression into the mating habits of women is this: long before the function of cells of the dividing embryo are determined, long before which sperm will find which egg is determined, molecules of our cell's surroundings are interacting with the cell's surface receptors to determine our behavior, and thus, the pool of sperm or egg from which we have to choose.

CHAPTER 4

The Sperm's Journey to the Egg

"Every sperm is sacred. Every sperm is great."

—MONTY PYTHON'S FLYING CIRCUS

Who We Are

I am sure you have taken the time to look at yourself closely in a mirror. It's something we've all done and still do, on some level. As you stare at your facial features you might recognize your dad's eyes and your mom's nose. From old photos you have seen you may notice a striking resemblance to your great grandmother. Maybe you look a lot like your brother Daryl and nothing like your other brother Daryl.[32]

I have two younger brothers. No, they are not both named Daryl. Each of us has a different father. With one of my brothers I share a strong resemblance; strangers can recognize us as brothers without being told. But my other brother I look nothing like; we don't share facial features or body type and a stranger would be hard pressed to guess we are related, let alone brothers.

Who we are physically, our looks, our body type, are based on the genetic material of our parents. But each parent is only going to contribute to 50% of that makeup. Furthermore, the contribution from each parent is only 50% of their genetic makeup. Two brothers with different fathers getting the same 50% from their mother might look very similar to each other while two brothers getting the exact opposite 50% might look nothing alike. Theoretically, one could argue this last set of brothers shares no genetic material.[33] Then, are they really related?

As my earlier discussion of Mary and Steve illustrates, our parents getting hooked-up is as much a cellular reaction to the environment, MHC molecules influencing the subconscious, as it is anything else. Our potential genetic compliment is now limited to these two people. But, that complement of DNA will be limited even further during the formation of our sex-cells and the subsequent fusion of such cells from these individuals.

Our sex-cells, the egg and the sperm, are two "half" cells that fuse to form a new cell line (the embryo). They are not really half cells; the egg has all the components of a cell that could immediately start dividing; i.e., machinery for protein synthesis, lots of mitochondria for energy production, fuel storage. The egg, even at this time, still has a full complement of DNA. On the other hand, the sperm, while being mostly a bullet of genetic material, also has many components found in most cells.[34] However, the sperm does have only half the genetic material of other body cells and it will not be synthesizing proteins. How any two of these specific cells comes together is very much a matter of cellular determinism.

A Sperm's Baggage

Let's consider how the sperm finds the egg. Upon ejaculation millions of sperm cells flood the vaginal cavity in their quest to reach the egg, not that much unlike fleet week in San Francisco, seamen everywhere. Of course, it is understood what sailors seek once they port after months at sea. Their success will depend on many variables: the manners, wisdom and knowledge given by their mother and commanding officers, the establishments visited and events attended, and the connection made once introduced. Not unlike these sailors the success of any one sperm cell will depend on the preparation with which it starts, the route taken, and its actions upon reaching the egg. Success for each, at each of these steps, is a matter of their make-up and surroundings.

The sperm that reaches the egg will do so in part because it starts with the right baggage. In this case baggage refers to the surface receptors the sperm cell carries. Several studies have demonstrated the presence of surface receptors seems to play a significant role in the sperm's ability to fertilize the egg. In a review of the subject, Rajesh K. Naz, states that more than 100 surface receptor proteins have been identified. Considering the various groups of receptors, cytokine, hormone and neurotransmitter,[35] tells us much of what the sperm may encounter on its journey; success in reaching and fertilizing the egg will be predicated on having the right combination of these receptors. Having the right combination of receptors will depend on events during formation and maturation.

Since sperm will not be synthesizing proteins, the surface receptors they have for the journey come from the process by which they are formed. Sperm form when special cells lining the tubules in the testis divide.[36] This cell division process serves to cut the number of chromosomes in half. This is an extremely important step. Remember, we are combining two cells at the end—sperm and egg—and so to maintain a consistent number of chromosomes we must half the original number before combining cells at fertilization.

The cell membrane of sperm cells comes from those cells lining the tubules of the testis. It is the surface receptors of these tubule cells that will determine how the sperm cells mature. These early sperm cells are stuck with the surface receptors they have at this point.[37] The membrane proteins of the sperm may change overtime, but it will be a result of the environmental molecules binding with surface receptors, not a response by the cells genetic machine interpreting its microenvironment.

The sperm cell's ticket has pretty much been punched; it has the surface receptors and internal molecules it's going to have the rest of its journey . . . mostly. Turns out there are specialized cells in the testis, nurse cells, which will influence the combination of receptors a sperm cell can have.

Nurse Cells

Sertoli cells serve as nurse cells in aiding the maturation of the developing sperm. These cells create compartmental spaces that serve to sequester the developing sperm cells from blood serum, a blood/sperm barrier.[38] Maturing sperm are in direct contact with Sertoli cells from start to finish.[39] Physical contact between cells like this is a result of surface receptors from each cell type binding to one another. The Sertoli cells serve to guide and direct the sperm cells on this initial stage of their journey. This is facilitated by each cell's ability to communicate by way of contact through surface receptors.

Sertoli cells secrete a number of glycoproteins that may also help the sperm survive later stages of their journey. One of these secretions is the hormone inhibin, which serve to prevent the early breakdown of the sperm's cap. This is an important protection because without it an early degrading of the cap will keep that sperm from fertilizing the egg. Whether or not a sperm cell will receive this protection has already been determined by whether or not it has the appropriate receptors to bind to inhibin.

The relationship between Sertoli cells and sperm cells is not entirely one way, nor should we expect it to be. In addition to inhibin, Sertoli cells secrete an androgen binding protein that binds to testosterone, thereby increasing the levels of testosterone in the center of the tubule. The levels of androgen binding protein secreted by Sertoli cells increases in the presence of sperm cells.[40] What we have then, are cells working in concert, through contact and molecular secretions, to help determine which sperm cell is adequately packaged.

In keeping with my sailor analogy, the success of any one sailor will depend on their "formation"—how they were raised, and their maturity level—wisdom gained from their commanding officer. When they set off in San Francisco it is with these "receptors" with which they have to work and with which their success will be determined. The road is treacherous, filled with dangers and distractions; it is only the well-equipped that succeed.

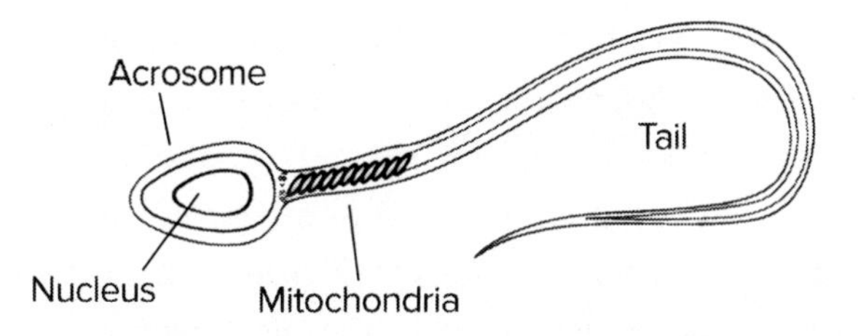

FIGURE 3: Sperm cell diagram. The acrosome protects the head until the cell reaches the egg, then it will degrade. The nucleus contains half the normal DNA complement. The mitochondria will provide the energy needed to whip the tail.

The Quest

An important part of any journey is your ride. For sperm that ride, their motility, comes by way of their long whiplike tail. This tail will play a significant role in the sperm's success and not surprisingly, we see various surface receptors influence this motility.

Even before ejaculation occurs, chemical messengers called cytokines, released by white blood cells, are influencing sperm motility. Sperm cells have been reported to contain various receptors for these chemical messengers.[41] Men suffering from spinal cord injury have elevated levels of cytokines in their semen; they also exhibit abnormally low sperm motility. Their production of sperm is adequate, but the sperm don't move like they should. However, when the actions of specific cytokines are inhibited, sperm motility improves.[42]

This kind of response when sperm cells encounter cytokines might be expected. The specific cytokines examined play various roles involved in our inflammatory response, fever and tumor destruction. High levels of such cytokines suggest the individual may be suffering from disease. With infection, inflammation and fever, the likelihood of any one sperm cell, or many sperm being malformed, increases. Encountering high levels of cytokines inhibits movement and will likely limit, if not end, the sperm's ability to reach the egg.

Evolutionarily, it's easy to see why this system might have evolved; sick, diseased men may produce a high rate of abnormal sperm you would not want necessarily creating the next generation. Again, we can imagine a sailor being prevented from disembarking because of an illness. When health improves and cytokine levels decline, then the healthy movement of the sperm returns. The microenvironment of the semen is determining which, if any, particular sperm will make it to the point of egg fertilization.

And We're Off

Once ejaculation occurs the sperm will find themselves in the hostile environment of the vagina. Many will succumb to the defenses of the female reproductive tract but those in the right position and with the right receptors will survive and swim forward. As we've seen, not having enough inhibin might cause an early cap breakdown and keep the sperm from reaching the egg. Additionally, as it turns out, not having enough receptors for the female hormone progesterone can also impact success.

It should not be surprising that we would find hormone receptors on the surface of sperm cells:[43] The environment the sperm cell develops in is bathed in hormones and the sperm will encounter hormones in the female reproductive tract. Progesterone's concentration in the female reproductive tract will be highest right after ovulation. Studies have demonstrated the presence of progesterone receptors on the surface of sperm and on the surface of both sperm precursors and Sertoli cells.[44]

Interestingly, it was observed that the concentration of progesterone receptors increases as the sperm develops and matures. In the early stages of sperm formation low levels of progesterone receptors were detected on immature sperm cells, but the intensity of progesterone receptors increases on more mature sperm cells.[45] The presence of these receptors in greater concentration as the sperm matures implies progesterone will play an important role in determining which sperm reaches the egg.

Such a role was recently demonstrated by Yuriy Kirichok and Polina Lishko in their lab at the University of San Francisco, "The effect of progesterone on sperm cells is very different than the effect it has on other cells" Kirichok states. He describes this effect as an electrical current that courses through the cell. This current has the effect of causing the sperm to begin to whip its tail in harder, larger strokes.[46] This whipping of the tail is called hyperactivation and is absolutely necessary if the sperm hopes to penetrate the egg.

Hyperactivation, however, doesn't begin until the sperm finds itself in close proximity to the egg. Prior to this, the sperm is swimming by small, fast tail movements that help propel it through the viscous fluids of the female reproductive tract. Because of the proximity to the ovaries and the attending cells of the egg, as the sperm approaches the egg the concentration of progesterone will increase. As more and more receptors of the sperm cell become bound by progesterone, the membrane potential of the cell changes and calcium ions rush into the cell, causing the tail to whip.

The action of the sperm changes with the concentration of progesterone molecules in its immediate environment. Fertilization usually occurs near the beginning of the uterine tube, so it would do no good for hyperactivation to occur in the vaginal canal or the uterus; it's still too far from the egg. It is likely the sperm is encountering progesterone molecules along the way but because the concentrations are not high enough, the tail whipping action is delayed. Imagine you are a sperm cell with 1000 receptors for progesterone. As you start your journey, you encounter a few progesterone molecules in the vagina; maybe ten. As you move into the uterus and up one of the oviducts, you encounter several more molecules; maybe 100. The concentration is increasing, but not high enough to get too excited. Then, you rush into an area where progesterone molecules are everywhere. Now, more than 500 of your receptors are bound by progesterone. That's an indication the egg is near and it's time to get busy; the action of the tail changes and the small, fast whips give way to the large, hard whips.

It has been suggested that these increasing concentrations of progesterone actually serve as a guide to the sperm. The sperm are detecting the increasing concentrations of progesterone and adjusting their behavior to move in a particular direction. This being the case, it is very easy to see how the sperm that will eventually fertilize the egg is partially determined by the amount of progesterone receptors it may have on its surface. Again, it's not that much different from a sailor making his way through the city in hopes of hooking up. And, of course, our sailor's behavior changes—more bravado maybe—in the presence of females.

The Destination

A third type of chemical messenger influencing the sperm's journey towards the egg is the neurotransmitter.[47] Generally, these are chemicals released by nerve cells to relay a message from one neuron to another. Neurons and sperm cells are two very different cells in many ways: nerve cells are stationary for the most part with many cellular extensions. Sperm cells are motile with no extensions save the tail. Nerve cells have a full complement of genetic material and are actively synthesizing proteins; sperm cells have half the amount of genetic material and are not synthesizing proteins. Yet, these two very functionally different cells have many of the same neurotransmitter receptors.[48]

One of the things that must happen if a sperm cell is to fertilize an egg is the process of capacitation and acrosome reaction (cap breakdown). Capacitation is a period in which molecules on the surface of the sperm (inhibin for instance, adsorbed while in the male reproductive tract) are removed by molecules found in the oviduct. Removal of these molecules allows for the acrosomal reaction to take place. Let's consider this point for a moment. While in the testis, specific molecules bind to the sperm cell as a means of guarding certain receptors from being bound too early, a function carried out by Sertoli cells. Once in the appropriate environment, these molecules are removed by environmental molecules to expose the receptors needed for the next stage. The sperm's behavior is being completely dictated by environmental molecules binding to surface receptors.

The acrosomal reaction is a process involving the fusion of the outer acrosomal membrane with the sperm membrane. The main purpose is to cause the release of the acrosomal content, which in turn will aid adherence to, and penetration of, the egg. This process of membrane fusion is also how nerve cells release their neurotransmitters into their microenvironment. Here, we have two very different cells, the nerve cell and the sperm cell, each relying on the same vital process to carry out their function. The nerve cell needs this to occur in order to communicate the electrical impulse to the neighboring neurons, and the sperm cell needs it in order to adhere to and penetrate the egg. If the process of membrane fusion is facilitated by the binding of neurotransmitters to surface receptors, then finding receptors for similar molecules on both cell types might be expected.

The next question then is, "Where does the sperm encounter these neurotransmitters that are going to determine which sperm makes it to the egg?" Studies have demonstrated the presence of neurotransmitters in seminal fluid and in the oviduct.[49],[50] Some neurotransmitters, like glycine, may have an inhibitory role when in the seminal fluid but as concentration increases the role may change to facilitate induction of membrane fusion.[51]

It's easy to see how a molecule could play a duel role. Imagine the number of surface receptors a sperm cell might have for any particular neurotransmitter. While still in the seminal fluid maybe only 20% of a specific receptor is bound, or maybe the combination of specific receptors, 20% of this one and 40% of that one, serve to inhibit certain cellular activity, in this case, acrosomal reaction and hyperactivity. If those concentrations change, which they are most certainly going to in the oviduct, you will see the behavior of the cell change. Now, 40% or 60% of a specific receptor is bound, or the combination of receptors changes from 20% for one and 40% for another to just the opposite. This percentage change in number of receptors bound causes a different response from the cell. Instead of being inhibited, which is the case in the seminal fluid, now, in the oviduct, the cell is experiencing significant activity and acrosomal reaction—both responses dictated by environmental condition and both responses suitable for location.

In keeping with our sailor analogy we can imagine, and I'll leave the details to the reader, that a lucky sailor, having disembarked and traversed the obstacles of the city, finds himself alone with a young lady. There may be external "membrane" shedding and ultimately fusion. I recognize my sailor analogy breaks down on several levels, and the behavior of sailors is no different than any other group of young men, generally. Still, superficially, the behavior of the organism is not that much different than the behavior of the cells.

Sperm cells have lots of different surface receptors. This really is no different than any other cell; all cells are covered with surface receptors. The interesting thing with sperm cells is the myriad receptors we see, their similarity to nerve cells, and the significant role they play in determining the sperm's path.

You could argue that any specific sperm reaching the egg is a random event: the concentration of chemicals in the environment, the positioning of any particular sperm cell, the luck that he is not wearing a condom. No doubt, there is a certain element of randomness in that any molecule binds to any surface receptor by the chance of the two coming together, but, the cell's response is not random. Each sperm is going to respond based on the number of receptors and the concentration of molecules in the environment. The cell's actions will be determined by these two factors.

The combination of hormone, cytokine and neurotransmitter receptors on the surface of a sperm cell, coupled with the concentration of molecules in the sperm's environment (a result of secretions from other cells) will determine the sperm's journey. We have no choice in which sperm reaches the egg, it happens as a matter of process, and that process is determined by cellular activity. Our genetic make-up is determined by the activity of our parent's cells during the time of copulation. Sorry for that image.

CHAPTER 5

The Egg

"So much of difference between a triumph and a flop is determined by . . . venue."

—ANATOLY BELILOVSKY

How Young Did You Say You Were?

I mentioned I have two brothers and that each of us has a different father. While I share many features with one, with the other I have little in common. This is not all that surprising considering the male contribution to each sibling in this case is from a different source. At best, getting all the same genes from our mother we would be 50% similar to each other and at worst, each of us getting the exact opposite set of genes we would have a 0% similarity. In reality it's probably closer to 25% genetic similarity between each of us. There is however, another variable that can contribute to the similarities and difference brothers might share. How old is the mother?

A man forms sperm from the time he reaches puberty until he dies. No sperm is very old though, with the average being around three days.

But such is not true for a woman's sex-cells. Her eggs will be formed long before she begins ovulating; arrested in a state of suspended division each egg is formed prior to birth. I am older than my brothers. My mother was 17 when she became pregnant with me. She and her eggs were very young. However, she was 27 when she became pregnant with my youngest brother, who, incidentally, is the one with which I share few physical similarities. Her eggs were much older. These eggs then, like my mother of 27, had been exposed to much more than the 17 year old egg I came from.

These ten years may have affected the eggs such that even though my brother and I share some genes, the environment may have turned off some of his so that it's as if we don't share them at all. I discuss this phenomenon more completely later but I mention it now because it underscores how our environment, even the environment of our mother's ovaries, affects who we are. And who we are is going to be determined by which egg, and when, gets fertilized.

"Come Up and See Me Sometime."

When one compares an egg cell to a sperm cell, except for their ultimate role of giving rise to the next generation, they could not be more different. The sperm cell is a small packet of genetic material covered with surface receptors that specifically serve to direct it towards the egg. The sperm cell will leave its host, if you will, and enter the hostile environment of the female reproductive tract. Many of these surface receptors will aid the sperm in avoiding female defenses while also guiding the sperm on its journey. The egg, on the other hand, has no such hostilities to defend against. The egg's journey will be comparatively short and assisted by many surrounding cells. Where the sperm "fights through thick and thin" to reach the egg, the egg is "coddled and nursed" as it prepares to receive company.

The egg is a large, non-motile, cell that contains essentially all the information and material necessary to direct the early development of the embryo. A maturing egg, unlike maturing sperm cells, carries out significant protein synthesis early on.[52] A developing egg has its entire genetic complement, and while no longer being able to replicate, it can still synthesize proteins. The egg is actively preparing molecules that will be used by the early cells of the embryo for division. Assisting the egg in its preparation to receive the sperm are special cells called follicle cells. These cells will aid the egg in its development and maturation prior to ovulation,[53] inhibit other follicle cells from developing,[54] and aid the sperm in finding the egg after ovulation.[55]

The process of egg formation contributes more than just genetic material; it also distinguishes the egg's behavior from that of the sperm's. The egg is providing the embryo with all the necessary molecules, structures and organelles. The sperm, on the other hand, is only delivering genetic material to what is essentially a fully functioning cell. Incidentally, this is why we can use mitochondria[56] to trace our maternal line. Mitochondria have their own small piece of DNA (a remnant from a once independent existence) and since only the egg contributes the mitochondrial organelles of the next generation, we can use their DNA to determine our maternal ancestry. If we shared no other DNA, at least my brothers and I would share mitochondrial DNA.

The male starts producing sperm cells at puberty and will continue to do so for the rest of his life; prior to this onset of production, sperm stem cells are quiet. It is only after an introduction of secondary hormones that these stem cells become active and start producing sperm cells. Females are quite different; they will have all of their immature egg cells even before they are born. The egg stem cells have a limited ability to divide and, by birth, most have either been converted to primary eggs or acquiesced. At birth, females will have approximately 400,000 primary eggs, of which only about 400 will develop further. After the initial growth phase of these 400,000 primary eggs ends, they will enter a stationary phase of suspended division; of which they will stay from birth until puberty.[57]

Then, upon receiving the signal form the attending follicular cells, some of these primary eggs will reinitiate division in preparation for ovulation. Many of these primary eggs will never mature and those that do can be anywhere from 10 to 50 years old. The final division of the egg does not take place until after ovulation and fusion with a sperm cell occurs.[58] Not only are these divisions delayed, but the partition of cellular material is unequal. With sperm cells, four cells of equal size and composition are formed. With egg cells all the cellular material stays with one cell while the other three presumptive cells become aborted polar bodies (small cells with little cytoplasm and half the genetic material of the original cell; much like a sperm cell without a tail). These polar bodies degenerate.

This process of egg formation means the possible pool of cells that can give rise to the next generation has already been determined long before Mary starts getting turned on by her co-workers MHC molecules. His sperm will be produced anew every few days with millions of cells potentially in the right position and with the right surface receptors (in actuality, only one is in the right position with the right receptors). In contrast, her eggs will only be produced once and the number of cells limited even before she is born.

Moreover, the determinative aspect goes even deeper than just the limited number of egg cells that will form; the genetic make-up of the egg has also been limited. Examining this division process and the formation of polar bodies, we see how "choices" have been reduced much further. When cells undergo the kind of division that only happens in the testis or ovaries, they first double their chromosome number then divide in half—twice. This works because chromosomes come in pairs. We have 23 pairs of chromosomes, and each new sex cell will end up with one member of each of these 23 pairs. After DNA replication takes place, each stem cell has 92 chromosomes or 46 pairs. The first division produces two daughter cells, each with one member of the original 23 pairs; however, these chromosomes are still attached to the replicated strand. A second division separates these two strands, forming two more daughter cells now with only one strand in each of the cells for each of the 23 pairs. This process of replicating the DNA, then undergoing two successive cell divisions, ends with four genetically unique cells.

This is true of sperm cells, but not for the egg. The stem cells that give rise to the primary eggs do not produce four genetically unique cells. Instead, as the egg matures, three of the would-be cells never form, becoming polar bodies. Instead of having four genetically distinct cells from which to "choose," women end up with one (Figure 4).

Our genetic options are limited by the process in which one of the four unique genetic sequences is determined. We really do not know how one sequence might be selected versus another; it likely has a lot to do with positioning. We suspect that how paired chromosomes line-up prior to division is random, so the distribution of chromosomal sets to one cell or another is also a matter of chance. But then one of these chromosomal sets will form a polar body and disintegrate and the other will remain to become the egg. This decision is likely determined by the concentrations of molecules in various places within the cell, as well as likely being influenced by the attending follicular cells. A high concentration of, say calcium,[59] in the south end of the egg could serve to favor the selection of the genetic sequence closest to it. Or, specific levels of specific proteins may induce cells with certain genetic sequences to become polar bodies. The unequal distribution of molecules within the egg determines the sequences of DNA that will be available to the next generation.

This unequal distribution of molecules within the egg is influenced by the attending follicular cells. These cells release hormones that have a direct influence on the egg: some will increase genetic activity, some will inhibit surrounding cells, and some will increase, or decrease, the permeability of the egg membrane to certain molecules. We would not expect all follicle cells to be acting exactly the same. Some may be secreting more or less hormone depending on how many surface receptors they have bound. Since the egg is surrounded by follicle cells secreting chemicals at different rates, we can imagine an internal cell environment where higher concentrations of certain molecules might exist in isolated locations. Such varying concentrations of molecules throughout the cell could easily influence which set of chromosomes

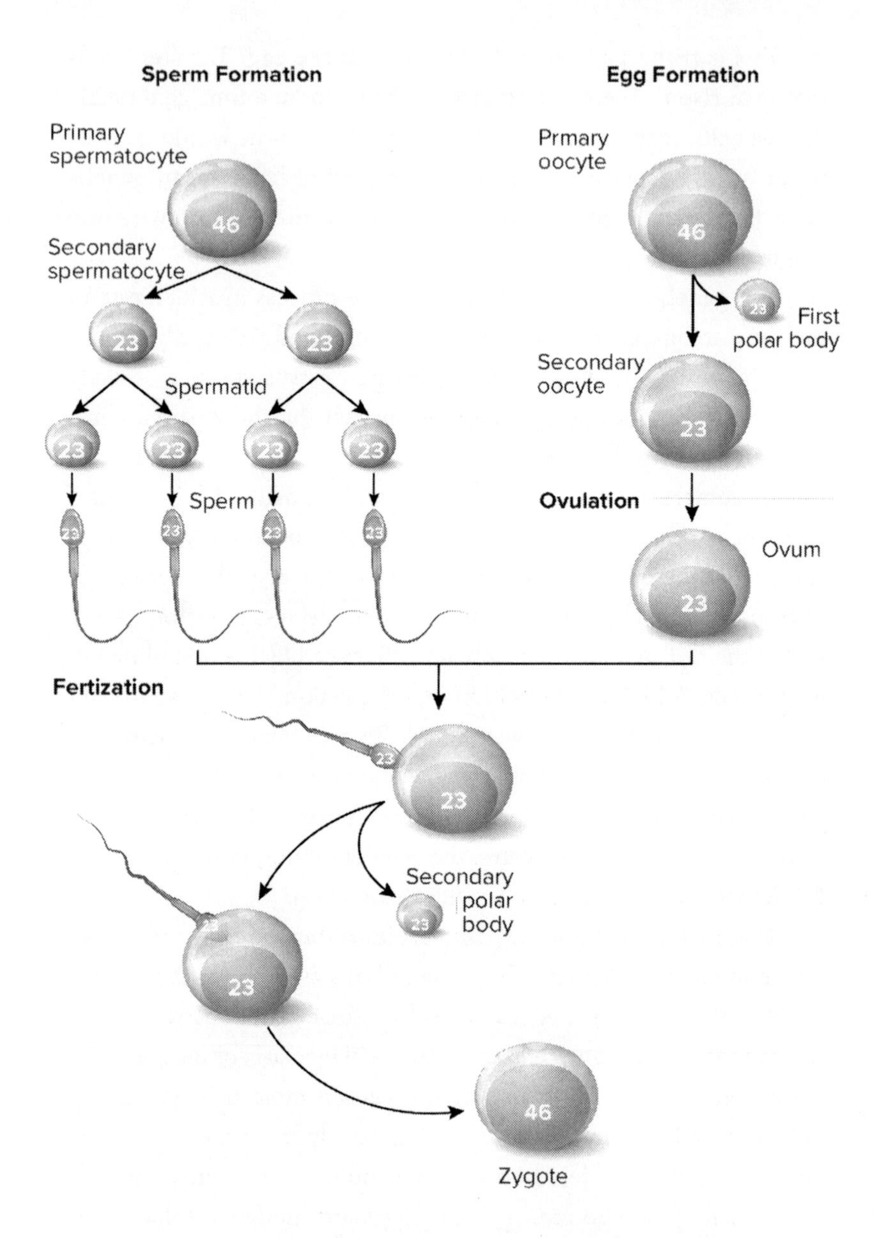

FIGURE 4: This diagram depicts the difference between sperm and egg formation. Spermatogenesis produces four equal haploid cells. Oogenesis produces one large cell and three polar bodies (the first polar body, if it continued meiosis would divide into two polar bodies—but it instead will degrade).

will become the mature egg and which set will be discarded as a polar body.

Follicular Cells

Let's take a closer look at these follicular cells. They initially surround the primary egg in a single, flat layer of cells. Their origin is from epithelial tissue of the ovary. These primordial follicles, of which there may be one million or more, will develop into about 400,000 primary follicles. These primary follicles, formed before birth, will remain quiescent until sexual maturity, and, for some, much, much longer. Then, at the age of sexual maturity, the body will release follicle stimulating hormone (FSH) which will make its way to the ovaries via the bloodstream to stimulate follicle development.

At puberty, all 400,000 or so primary eggs are ready to begin maturation. Obviously, we wouldn't want all eggs to mature at the same time (it would certainly create a very narrow window in which we could reproduce, as well as, a very large "litter"). Thankfully, they don't.

With the onset of sexual maturity the bloodstream delivers a flood of FSH to the cells of the ovary; however, the cells don't all get the same amount FSH at the same time. It takes time for the FSH to diffuse from the bloodstream to the follicle cells surrounding the eggs. So, of course, those cells closest to the blood supply are likely to encounter FSH molecules first. Upon binding to the surface receptors of the follicle cells, FSH will stimulate these cells to start to proliferate, and as they proliferate they will also begin to actively secrete hormones. These hormones will influence the eggs maturation as well as the development of follicle cells that might be nearby.

As FSH floods the ovaries, several primary eggs begin to mature at the same time, but as the follicular cells continue to grow, the hormone concentrations they are releasing begins to increase. One of these hormones is inhibin (the same hormone released by Sertoli cells), which serves to inhibit the action of other follicle cells.[60] One or two of the developing eggs will become dominant;[61] its secretions serving to further enhance its own development while simultaneously serving to inhibit the development of other eggs.[62] In this way, the amount of eggs that might mature at any one time is limited.

Which egg is "chosen" to complete maturation to the point of ovulation? Again, it is likely due to position and surface receptor concentrations. While several primary eggs may begin maturation because of the FSH signals, one will eventually take the lead. Imagine that as the FSH molecules move in, there are going to be those follicle cells that will encounter them first and so, therefore, begin responding first. They will start proliferating and they will start secreting their hormones. As they grow, their secretions will increase. Inhibin will interfere with the continued release of FSH, reducing the number of bound FSH molecules on the surface receptors of other follicle cells, thus serving to slowdown and eventually halt the development of other eggs. It really becomes a race between developing primary eggs as to which one will eventually reach maturity. Eventually, one primary egg will be producing enough hormones that all other developing primary eggs will acquiesce and eventually undergo degradation (atresia). The egg that matures to ovulation in any given cycle is determined by its relative position in the ovary and the direct response of its follicular cells to the presence of FSH. (Figure 5)

In young women, where there are hundreds of thousands of potential primary eggs available to respond to FSH release, we see that many primary eggs begin maturation only to be shut down and eventually degraded. This means as a woman ages, the genetic options, that is, the genetic variability she has to offer the next generation, is reduced by each egg that undergoes atresia. In fact, estimates are that by the time she is thirty she will only have about 10% of her initial primary egg concentration remaining.[63] Whereas a man can produce sperm with any possible genetic variation well into his very senior years, the genetic variability a woman can offer the next generation diminishes with each ovulation. The biological clock women feel ticking is actually an hourglass.

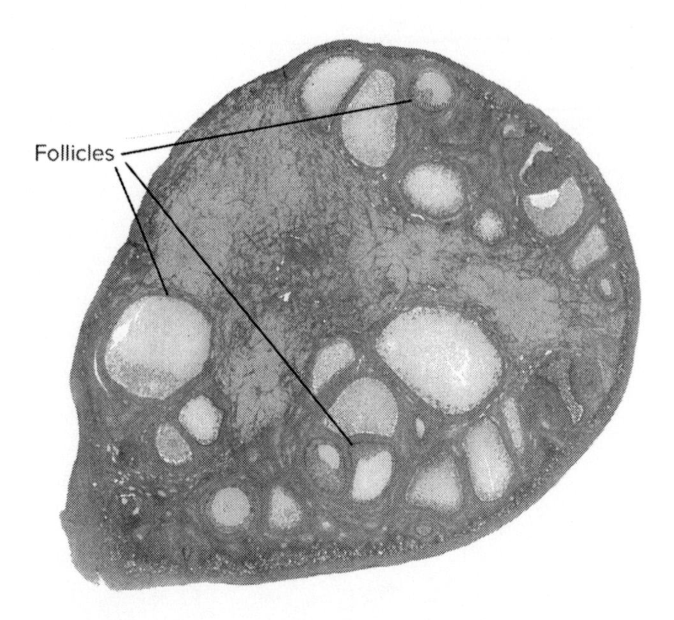

FIGURE 5: Cross-section of an ovary showing several follicles at various stages of maturity.

The selection of the egg, as with the sperm, is a matter of cellular activity in response to environmental conditions, micro-environmental conditions. The cells can only respond to those molecules in their immediate surroundings. These molecules can come from cells near (follicle cells) and far (FSH from the pituitary). The next generation is being determined by the interaction of cells, egg and sperm in this case, to their local environment; which, in turn, is being determined by the cellular response of the organism to its macro-environment.

CHAPTER 6

Growing and Development

"Potential has a shelf life."

—MARGARET ATWOOD

Such Potential

I have always told my daughter she can be anything she wants. Most parents do the same, having high hopes for their children, they will tell them with hard work and effort they can become anything, even president. But, every parent also realizes their children have limitations. These limitations become more evident as a child ages. My daughter is an extremely bright girl, but it became evident to her mother and me on those car trips to grandma's that she was not going to be a singer. It is also true to say I knew she wasn't going to be a linebacker. You know fairly early on if your child has any athletic ability, or artistic talent, or is exceptionally bright, or not these things. We begin life with great potential, but this potential has already been limited by the physical qualities and characteristics we inherit.

Those limits are compounded when we consider the environment our children are exposed to during development will also influence their potential. A great athlete doesn't become one if he never gets to play sports. Great singers and artists and scholars must have an environment conducive to their skill development. Our potential, as great as it is, is limited by who we are and what we are exposed to.

In The Beginning

We begin life as a single cell, a zygote (fertilized egg), that divides shortly after fertilization and continues dividing until our death. That's not to say all cells of the organism continue to divide—most cells will eventually undergo a process of determination in which the cell's function, and possibly number of divisions becomes fixed. Still, there will always be some cells dividing: epithelial cells, blood stem cells, spermatocytes. Many cells will die, some will be replaced and more will carry out their task until the death of the organism. All cells, however, are derived from this same, initial, single cell and all have the same genetic make-up. So then, what causes one cell to become a muscle cell and another to become a nerve cell? The process of embryonic determination—how our cells know what to become and what to do—is a matter of the cell's microenvironment.

The zygote, resulting from the fusion of egg and sperm, is a new genetic compilation. Because of crossing-over, independent assortment,[64] etc., that occur during the process that leads to egg and sperm, this specific genetic combination likely has never existed before; a new genetic line with the potential to live forever, as we saw with the cells of Henrietta Lacks. The sheer existence of this particular genetic line has been determined by the cellular biology of two different organisms, both responding to their own cellular activity. This new chromosomal arrangement; this new grouping of genes; this new sequence of nucleotides, has the potential to be the greatest organism to ever exist. This particular DNA sequence might code for one of the greatest scientists, or greatest athletes, or greatest chefs, all based on the types of proteins synthesized. The proteins being synthesized, directly from the genes of these chromosomes, will determine the type, range, and number of surface receptors the cells of this organism will have. This in turn will determine each cell's activity, which ultimately determines what the organism will become.

But just having the genes is not enough—the cell will have to interact with the environment. This interaction with molecules of the cell's surroundings can greatly influence how these cells, and the organism, develops. Even before this new genetic sequence begins dividing, it is being influenced by the mother's cellular biology. Fertilization likely has taken place in the oviduct and the microenvironment of the zygote is flooded with hormones from the mother. The mother's early environment will play a pivotal role, not just now, but throughout development. Actually, as will be discussed later, her environment, and maybe even her grandparent's environment, can determine which genes may or may not be turned on or off. For the rest of the organism's life this early exposure to its mother's chemical biology will dictate what and how the organism's cells, and ultimately the organism itself, react to its environment.

Embryonic Development

After fertilization, the egg will complete the meiotic reduction process, leaving it with the 23 chromosomes that will fuse with the sperm's 23 chromosomes to form the nucleus of the new zygote. Shortly thereafter, the zygote will begin its first mitotic division (DNA is replicated and the cell pinches in two by a process called cleavage). Two more divisions will give us a ball of eight cells. These early divisions are basically splitting up the cytoplasm of the egg into more manageable volumes—there is no increase in size; the cells replicate their DNA, but they do not grow before dividing. At this time, all these cells are said to be totipotent: they can become any type of cell. This differs a little from pluripotentcy in which cells can become any type of body cell; totipotentcy includes the ability to become a cell of the extraembryonic membranes (placenta). Theoretically then, if we could separate this ball into eight individual cells we would get eight identical people.

While it is true these cells are undifferentiated and have the ability to become any type of cell, I think it would be wrong to consider them completely undetermined. The cell's fate, what it will become, has already been determined by its position. Norton B. Gilula demonstrated that cytoplasmic bridges form between sister blastomeres (early cells of the developing embryo) at the two, four, and eight-cell embryo.[65] The cells next to each other exchange cytoplasmic molecules by aligning small openings in the cell membrane that then permit the exchange of ions like NA+, K+, Cl-. These pathways have been found to also be permeable to other low weight molecules.[66] These ion channels permit communication between neighboring cells, but indirectly limit communication between more distant neighbors.

At the eight cell stage the embryo undergoes a process called compaction: in this phase, the cells bind tightly together, increasing their surface contact with neighboring cells. Gilula demonstrated that junction-mediated communication occurs between cells during compaction. He was able to observe injected fluorescein transfer by way of ionic coupling. That is, fluorescein moved from one sister blastomere to another sister blastomere through these small passage-ways between cells. This communication was retained throughout compaction and well into the next developmental stage: fluorescein was observed to spread from trophoblast cells to other trophoblast cells, as well as cells of the inner cell mass.[67, 68] This increased surface contact caused by compaction allows for greater communication between neighboring cells; this, in turn, determines what each of these eight cells will give rise to. While these eight cells are totipotent—only if separated from the others; if left where they are, their fate has been determined. There are many cell types the mitotic descendants of each of these cells cannot become.

Imagine these eight cells jammed together, sharing as much surface area as possible: each cell relaying its position to the others, while also determining its position in relation to the others. The cells each is in contact with are going to influence what each cell becomes.

Even after the very first division, we have two cells with different fates. In 1980 Bennett Shapiro and his team demonstrated that parts of the sperm cell end up in one of the first two cells.[69] They labeled the surface of mouse sperm with a fluorescence dye and then followed the movement after fertilization. It had been assumed that the sperm, after fusion, would be degraded. Instead, Shapiro demonstrated that much of the sperm surface components were persistent in the embryo throughout its earliest development. It is not known if the presence of the sperm's surface components has any influence on how the cell might develop or function, but it does create a difference between the two cells.

Additionally, the environment of each cell after the initial division is different simply because one is on the right and one is on the left. Of these first two cells, each is experiencing slightly different environmental molecules and concentrations because of position. So, as it is for all the cells of our initial compact ball of eight cells. We describe these cells as totipotent, but that is only true if we manipulate their environment. If we leave them alone and let development take its course, then the fate of these cells, with regards to what they will become, was determined the moment they were formed.

Cell Differentiation

Compaction, which leads to a polarization of the cells within this solid ball, is followed by two more divisions to produce a 32-cell-staged embryo. At this point, we see the cells begin to realign themselves as division continues; some cells move to the periphery of our ball and some cells remain inside a cavity that is being created by the movement of the cells. This process, called blastulation, creates a hollow ball of cells. A single layer of cells forms the outer surface of our developing embryo; these cells are trophoblast cells and will take part in the formation of the placenta. Trophoblast cells that, just two divisions earlier, had the potential to be a fully functional individual, are now relegated to being supporting cells with a limited life of about nine months. The directive came during compaction while the cells were communicating. Chemical composition and position dictated, "You cells will divide and give rise to the placenta."

While true the cells of the solid ball phase are all totipotent, with the potential to become any other cell type, it is also not true. We still have to recognize that these cells are going to continue to be influenced by their environment, both by the cells they are in contact with and by the molecules of their microenvironment. So, while they do still have totipotentcy, that is only if something in the environment changes, i.e., artificial separation of cells. If development continues normally, then the cells that make up the trophoblast were determined well before they began to differentiate.

Lying just inside the outer sphere of trophoblastic cells, but still attached, is a small group of cells called the inner cell mass (Figure 6). This group of cells, about 12 initially, is also known as the embryoblast. It is these cells that give rise to the individual. Yes, it is only a small number of the original cells that will actually develop into the organism.

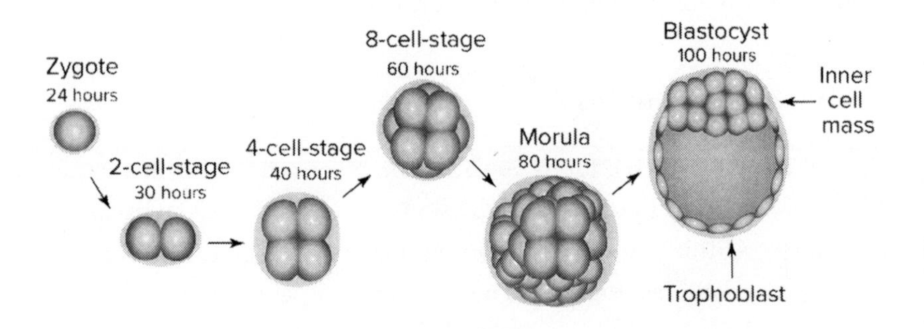

FIGURE 6: The first several stages of embyonic development. Note the mass of cells forming at the anterior end of the blastocyst, this is the inner cell mass that will give rise to the individual. The ring of cells around the periphery are the trophoblastic cells that will form the placenta.

The cells of the inner cell mass are pluripotent: They have the ability to develop into any cell type, except, trophoblast cells. There is a slight difference between being totipotent and pluripotent. Still, as demonstrated by Gilula using fluorescein,[70] there is communication going on between trophoblastic cells and the inner mass cells, but it is declining. The cells of both the inner cell mass and those to become the placenta have received their orders, exchanged information to make sure everyone's on the same page and set out to implement the plan. Of course, each cell has the flexibility to modify its response depending on how the plan unfolds.

This flexibility is demonstrated with the production of identical twins. While the mechanism is not completely understood, in 2007 Dr. Dianna Payne [71] observed a collapsing of the blastocyst to be the cause of identical twins. For reasons not clear, about three in a thousand embryos collapse. This causes the cells of the inner mass to be split in half. Essentially, forming two groups of progenitor cells on opposite sides of the embryo. Now the cells of the inner mass that were communicating and preparing to differentiate further, find their environment has changed. Some cells that were in direct contact with other cells have now been separated. They are no longer receiving data from those cells. This will result in the cells both receiving a different message, and upon interpretation, transmitting a different message. While a 12-celled, inner cell mass is ready for further development, a 6-celled mass will have to undergo another division before it is ready. These changes in environment are detected by the cells and they adjust accordingly. The end result is two individuals that are genetically identical.

These two individuals, twins, derived from the fusion of the same egg and sperm have all of the same genetic information. It's not unreasonable to think they would develop identically as well, but they do not. It is true it can be very difficult to tell identical twins apart; they look alike, their mannerisms can be the same; yet, they are different. The environment at the cellular level will be different for one inner cell mass than for the other and this difference will influence how the cells of each cell mass will respond.

Consider fingerprints. Identical twins do not have the same fingerprints.[72] The difference is largely due to the amount of blood flow each fetus receives; which, in part, is determined by the length and diameter of the umbilical cord.[73] So, while we might not be able to tell them apart genetically, it turns out the environment allows us to tell them apart phenotypically.

As the inner cell mass grows, the cells will become more and more specialized. They will lose their pluripotentcy as they respond to the cells around them. First, they will be divided into cells of different germ layers: ectoderm, mesoderm, endoderm, each giving rise to specific types of cells. The ectoderm gives rise to both skin cells and nerves cells, the endoderm the digestive lining, and the mesoderm the internal structures. The germ layers will then differentiate into tissues: muscle, epithelial, connective, nervous. Within these tissues, the cells will differentiate into very specific, specialized cells, carrying out a function that is being determined by neighboring cells.

During the entire process of embryogenesis through organogenesis (basically, from the first division through formation of organs) the fate of the cell is being determined by the microenvironment surrounding it. That microenvironment is a matter of cell-to-cell contact and those molecules able to attach to the cell's surface receptors. These pieces of information are relayed back to the nucleus and, either by turning on genes or turning off genes, the cell adjusts its protein synthesis to meet the needs necessary to carry out the function being dictated by the cell's surroundings.

Imagine, at some point during development, the ectoderm will specialize into cells that will become nerve cells and cells that will become skin cells. This means at some location in the embryo two cells right next to each other are destined for two very different functions. Being right next to each other, you would expect that their surroundings would be very similar; that is, the molecules in the microenvironment should not vary much considering the cells are right next to each other. So, what determines which one becomes nerve and which one becomes skin? In this case, the surrounding cells play a significant role. Even though these two cells are in contact on one side, the other sides are likely also interacting with different cells. If the cell on the right is interacting with other cells that are becoming nerve cells, this influence may have already resulted in different surface receptors that lead it to become a nerve cell. On the other hand, the cell on the left is surrounded by other destined skins cells and so the information received from these cells directs it to produce receptor proteins reminiscent of skin cells. Even though these cells may be awash with similar molecules, the influence of neighboring cells has resulted in different surface receptors and, therefore, very different cells with very different functions.

This is actually borne out by several studies conducted by H. Spemann in the 1920s.[74] In numerous experiments, Spemann transplanted small pieces of presumptive skin ectoderm into presumptive neural plate areas and the transplanted cells developed in accordance to the surrounding cells. When he went the other direction, transplanting cells of neural plate to presumptive skin ectoderm, the cells again developed in accordance with the cells around them. This was also true if he transplanted cells of different germ layers; they would develop in accordance to their surroundings. However, this was only the case if the transplants were made in early gastrulation (a period of cell movement during the formation of germ layers). When Spemann transplanted cells of the late gastrula the results were quite the opposite. Neural plate cells from late gastrula, transplanted into presumptive ectoderm, will continue to develop into nerve cells. Presumptive skin cells from the late gastrula stage, transplanted into neural plate tissue in late gastrula will continue to develop into skin tissue.[75] At some point the cells fate is fixed and its pliability lost.

Why? Because of the proteins and surface receptors being produced by the cell. In the early stages of the gastrula, the cell's surface receptors are highly varied, the protein production has not yet been specialized, but as the cell's time surrounded by other like cells increases, so do the surface receptors and proteins the cell is producing to aid this communication. At some point, in the late gastrula it appears, the cells surface receptors and protein production are set to respond to becoming a nerve cell. Moving it to a new location does not affect its development because the cell no longer has the surface receptors to allow it to recognize what these new, surrounding cells are saying. "What do you mean become a skin cell? I'm sorry, I can only understand neuraleese." This is exactly why it continues to become a nerve cell. Its receptors are going to respond to molecules, whether as abundant or not, that influence neuronal development simply because those are the receptors the cell has.

As development continues, from cleavage on through the morula and the blastocyst, into and through organogenesis, the cells become more and more specialized. The loss of pliability is a result of the cells responding to their environment as the embryo grows. The cells behavior is determined by its environment and its ability to respond to that environment. The environment directs and dictates the receptors and proteins the cell will have and produce, but ultimately, only in accordance with the genetic capability of the cell.

CHAPTER 7

Our Blueprints

"Always remember that you
are absolutely unique.
Just like everyone else."

—MARGARET MEAD

Genetics

Much of the discussion so far has focused on how the environment and, more importantly, the microenvironment, of the cell influences the cell's behavior; such a discussion reflects the nurture side of the argument. However, we do not want to sell short the influence genes have on how we respond to our external stimuli. There really should be no nature versus nurture argument; they (nature and nurture) are one; each working in accordance with the other. Without nature (the cells), nurture (the environment) has nothing to influence, and without nurture, nature has nothing to which to respond. How do genes then manifest their influence on our behavior?

DNA is the raw material of which we are made. Well, not so much the raw material, there are a number of other molecules that go into the making of a cell, as much as the raw instructions. These instructions, while quite rigid (the genetic sequence established when egg and sperm fuse, is the sequence all your cells will have, except for the occasional mutation, for the rest of your life) are very open to interpretation. Analogously, just like any book read by two people, the impression and opinion of what was read, while it may be similar, will still be different between the two individuals. Consider the Bible. The story has been the same for hundreds of years and read by millions of people, and every one of those individuals has had their own idea of what the words mean. Our genetic code is very much a book being read by the environment. The environment can only read the words of the book, it may read them selectively, but it cannot add words (there is an exception here to be explained shortly). We should not think or talk about DNA initiating anything; whatever gets transcribed from the genes is a result of the cell's environment. The book does not tell the reader to read chapter seven, regardless, the reader may read *only* chapter seven.

In this analogy, like nature versus nurture, we have to realize book and reader exist together. Without the reader, the words of the book cannot be interpreted, and without the book, there is nothing for the reader to interpret. Each cell has a set of instructions and cells of the same organism all have the same instructions, but each of these sets of instructions will be read slightly differently by the environment. This may not a perfect analogy, but close enough in that the DNA sequence is only the words to be interpreted by the surroundings of the cell.

How then are these words, the genes, expressed at the cellular level? The sequence of DNA to be transcribed is copied into messenger RNA (mRNA); this molecule, after some modification, strings together amino acids using the nucleotide sequence copied from the DNA molecule to determine the correct order. The order of amino acids is very important because it will dictate the shape of the molecule being formed; in the case of these molecules, shape is everything. Each gene (a specific segment of DNA) codes for a specific sequence of amino acids. As the mRNA molecule strings together the correct sequence of amino acids, the string will fold into a particular shape, producing a large macromolecule with a very specific size and shape. This macromolecule is a protein. Protein molecules are the language of our genes. If the DNA is the written word, then protein molecules are the spoken word.

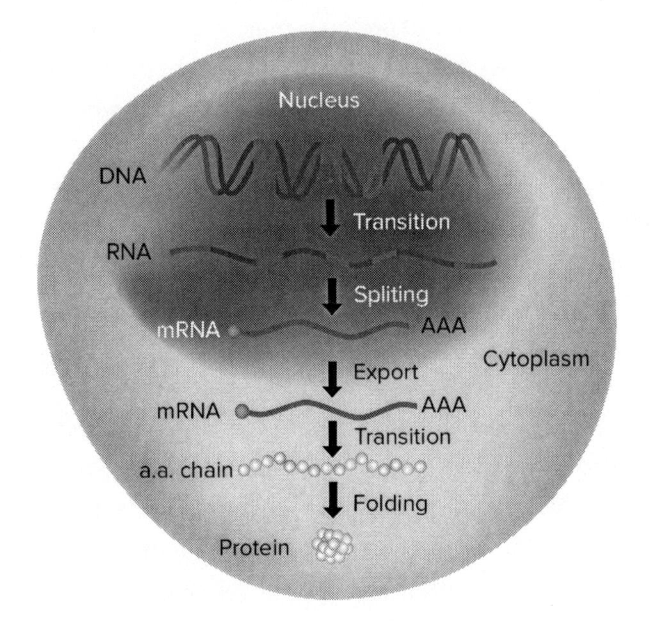

FIGURE 7: The sequence of events from DNA molecule to protein.

These proteins come in two fundamental forms, either structural, or enzymatic (catalystic). Structural proteins are used for building the different parts of the cell, we can think of them as the wood and bricks. Enzymatic proteins can be thought of as the tools used in the building and operation of the cell. Enzymes cut and shape larger molecules much like the saw and hammer at a construction site. But enzymes are highly specific, as well. They will only work on specific molecules. This is why their shape is so important; if they are the wrong shape, they will not be able to function. In the case of enzymes, you must use the right tool for the job.

Proteins, both structural and enzymatic, and found both within the cell and embedded in the cell membrane. It is these proteins with which the environment is interacting, and it is this interaction between environment and protein that determines how the cell behaves. It actually becomes a vicious cycle—the environment interacts with proteins, that, in turn, dictate which genes will be read, determining which proteins the cell will produce, thereby establishing the surface receptors (proteins) with which the environment can interact. Genes provide the proteins that interpret the environment, which leads to a response from the cell—the cell's behavior.

If the gene is not present then the environment cannot influence the cell on that specific level. As important and influential as the environment is to the behavior of the cell, it is still going to be limited by the available proteins with which to react. The Human Immunodeficiency Virus (HIV) is a prime example of the environment's dependence on the cell's receptors. HIV must bind to two receptors on the surface of the cells it infects. One of these is a co-receptor that is estimated to be absent in about 10% of the Caucasian population.[76] Because of this genetic defect, these individuals do not code for this protein, and so HIV cannot infect those cells. In this case, missing or malfunctioning genes limit the environment's impact on the cell.

Adding to the Book

Earlier, I mentioned the environment, like the reader, cannot add words to the book; well, that's not entirely true. It is true, in general, our cells all have the same DNA sequence we started with so many years ago—this is why forensic science works. However, it is quite probable that some of our cells have extra DNA we didn't have when we were born. The extra DNA is picked up from our environment and can have devastating effects on the cell and ultimately the organism. Where does this extra DNA come from? Viruses. Not all viruses, but many viruses, when they infect our cells will integrate their own genetic material into our DNA. Examples of such viruses include herpes, HIV, and chickenpox. If you have, or have had, any of these diseases, the genetic material from those viruses still resides in your cells. It is in this way the environment can add words to the book.

In fact, evidence now shows as much as 100,000 fragments of our current DNA are actually residual genetic material from bygone viral infections.[77] Viruses like HIV and herpes are known as retroviruses; these viruses infect a cell, then using viral enzymes they bring with them, they insert their genetic material into the genetic material of the host cell. The genome of the host cell now contains a nucleotide sequence belonging to the virus. On occasion, the virus may fail to get "turned on" and so does not produce more viruses. The viral DNA becomes "trapped" in the host DNA. When this type of infection happens in egg or sperm cells, the genes of the virus can be passed to the next generation. In this way genes that were only in a few cells initially, egg or sperm, will be in every cell of the individual in the next generation. This process seems to have been going on for millions of years as best estimates have our genome currently being 8% viral;[78] that number may rise.

Viral segments tucked into our genome are called fossil viruses and until recently they have all been retroviruses. However, in January of 2010 a team of Japanese scientists reported the presence of four segments of human DNA that were from a borna virus.[79] Borna viruses are not retroviruses and so they do not integrate their DNA into the host. So, how did this viral DNA get into our DNA? Although borna viruses do not integrate into the host DNA, they do reside in the nucleus of the cell in close proximity to chromosomes. "Mammal genomes contain thousands of stretches of DNA . . . [which] sometimes make copies of themselves that get reinserted back into the genome."[80] Scientists suggest one of these repeating segments could have captured nearby borna virus DNA as it was reinserting itself, thus adding it to the host sequence of nucleotides.[81] It is likely this process has occurred many times.

Do we synthesize any proteins using viral genes leftover from ancestral infections? As it turns out, we do. A team of virologists has identified the protein syncytin, secreted only by cells involved in placental formation, as being viral based.[82] Syncytin is found in the cells that form the single cellular layer fusing the placenta with the uterus. Fusion of cells seems to be the original function of this viral gene; it served to fuse host cells together, allowing the virus to spread from one cell to another. Today, this fusion is necessary for nutrients to be passed from mother to fetus.[83]

This team of virologists has demonstrated, using mice as subjects, that when you shut down the syncytin gene, the embryo dies at day eleven because the syncytiotrophoblast (fusion of uterus and placenta) does not form. The gene that codes for this apparently critical protein is the result of an acquired viral infection early in the evolution of placental mammals. They have identified at least six different syncytin proteins (from primates, rabbits, dogs, and cats) likely from at least three different viruses, actively being synthesized.[84]

This type of characteristic acquisition, production of new proteins as a result of acquired genes, which then get passed to the next generation, is Lamarckian evolution. Jean-Baptiste Lamarck was a French naturalist who, in 1800 during a lecture at the National Museum of Natural History in Paris, proposed the idea of inheritance through acquired characteristics. Long before Darwin, Lamarck was suggesting "change in species occurred over time."[85] The pressure for this change was environmental and the principle means by which change occurred was through use/disuse models. Something not used would be lost in the next generation while something used would grow stronger and be passed on. For example, blindness in moles, because they were underground creatures who had no need for seeing, their eyesight was lost through disuse. Lamarck was not exactly correct with the mechanism, but as we have seen with fossil viruses, and as we will see with the next section on epigenetics, the idea of organisms acquiring traits as a result of their environment, then passing these acquired traits to their offspring, has been confirmed.

Epigenetics

It is not unusual for students to give me material: articles, books, videos, they think I will find interesting. In February of 2010, a young lady gave me a video of the Nova production, *Ghost in our Genes*. When students give me stuff I will do a cursory review of the material to be polite, but I generally do not have the time to read or watch everything. However, this particular young woman had already demonstrated an attentiveness in class, and a tenaciousness in seeking answers, that I knew, would not accept a cursory review. Knowing she would expect more than a superficial discussion on the video, I sat down to watch.

I will not tell you I was blown away by the material, but what the video did was reinforce this concept I had been considering; organismal behavior was a matter of cellular behavior in response to the environment. In this particular case, the environment was influencing which genes were on and which genes were off. To me, this made perfect sense; of course what was happening in the cells' environment should affect which genes would be expressed—is that not the whole premise of cell differentiation during development? What was remarkable was the suggestion that the switching on or off of genes in one generation, due to environmental influences, could be passed to future generations.

The process of switching genes on or off is called epigenetics. The term means "on top of the genes" and refers to functional changes in expression of genes without changing the underlying nucleotide sequence. Closely associated with the DNA molecule are proteins called histones. These specialized molecules serve to keep the DNA in ordered, packaged units called nucleosomes. Histones function much like a spool with the DNA molecule neatly coiled around it. In a process called methylation, methyl groups can bind to the histone proteins and/or the DNA molecule. When this happens, the DNA molecule becomes more tightly bound to the histone proteins.

A nucleotide sequence that is tightly bound is difficult to transcribe; this serves to prevent the expression of that gene.[86] (Figure 8)

How would methylation then get passed on to the next generation? One method of methylation of the DNA molecule occurs when a methyl group is added to cytosine (one of the four nucleotides composing DNA) forming 5-methylcytosine. It's important to note 5-methylctosine will still pair with its complimentary nucleotide,

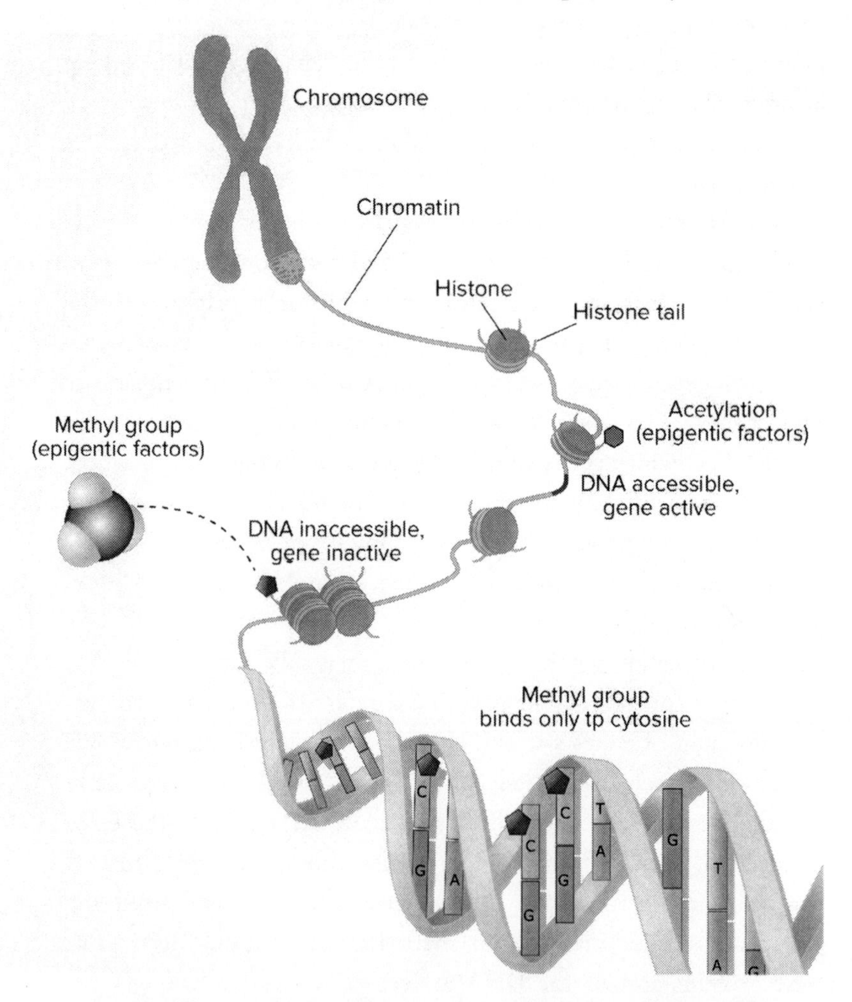

FIGURE 8: The epigentic mechanism; as a means of keeping the DNA orderly, it will coil around histone molecules, when these become bound by methyl groups the gene cannot be transcribed and is essentially shut-off.

guanine. Certain enzymes have the tendency, if they come across a segment of DNA with which only one strand is methylated, to methylate the other strand as well.[87, 88] When the DNA molecule is replicated, the methylation of the molecule is conserved and the suppression of the gene passed to the new cell.[89] In this way, a gene could be switched off in one generation because of environmental conditions and then the lack of expression for that gene passed on to the next generation—Lamarckian evolution.

A Mother's Love

Mice, and mammals generally, have a gene (*agouti*) responsible for fur color and, at least in mice, a tendency towards obesity. When the gene is turned on, the mice have a yellow coat and eat excessively. The protein from the gene interferes with receptors in the brain that tell the mice they are full and so the mice eat themselves into obesity. In 2002, Craig Cooney and team demonstrated that the expression of this gene in the next generation could be manipulated by altering the diet of the mother.[90]

Using two genetically identical strains of mice, the researchers fed them one of three diets (control, mid-range, and high-level) of methylated supplements (vitamin B12). The diet had a huge impact on the mice. Fat, yellow mothers that were fed diets high in methyl supplements were giving birth to skinny brown offspring. In fact, as levels of methyl supplementation increased, so did the distribution towards a more agouti coat (increasing combination of black and yellow pigment in the hair). Apparently, the agouti gene became methylated as a result of the increased levels of vitamin B12 in the mother's diet. The methylation of the gene served to inhibit its transcription—shutting it off. With the gene shut down, the protein that interfered with the receptors of the brain used in measuring satiation was no longer present. The brain cells of these mice can now receive the message they are full and stop eating. Thus, the environment of one generation has had a profound impact on the phenotype of the next generation: the 'behavior,' eating methylated supplements, of one generation serving to determine the cellular activity, not producing satiation blocking proteins, of the next generation.

There is evidence that not only diet, but experience can also affect the epigenome. Researchers have shown that the conditions rats are raised under can also greatly influence gene expression.[91] Like people, mice have a variety of nurturing behaviors. Some mother rats are strong nurturers, licking and grooming their offspring intensely after birth; other mothers can be less nurturing, taking a more distant approach and licking the offspring much less than nurturing mothers do.

When researchers looked at the offspring born to these two types of mothers, they found that under stressful conditions the offspring of less nurturing mothers, the low-lickers, were prone to greater increases in blood pressure and stress hormone production. When they switched the rats so that the offspring of high nurturing mothers were given to low-lickers, and offspring of low nurturing mothers where given to high-lickers, the results did not change; those reared by low-nurturing mothers responded poorly to stress despite being offspring of high-nurturing mothers. Clearly, the different responses were not the result of genes, per se, but the result of the mother's behavior.

Somehow the mother's behavior had affected gene expression.

The researchers next examined the gene responsible for lowering the levels of stress hormone in the blood. This particular hormone is most active in the hippocampal region of the brain. Comparing the hippocampus of rats from high-licking mothers with those from low-licking mothers provided striking evidence of the epigenomic influence on gene expression. The "less nurtured rats had multiple epigenetic marks silencing the gene."[92] The gene responsible for lowering stress levels is less active in rats that received less nurturing, and subsequently, their stress levels soared.

Here again, we see a condition where the cellular activity of the offspring is directly influenced by the behavior of the mother. The environment, this time the mother's behavior, is having a direct impact on the types of proteins the offspring will synthesize; ultimately, one's behavior then is a result of cellular activity, in part, determined by the behavior of another.

We have all seen parents, maybe we know some (maybe we are some), who seem to take a hands-off approach to raising their children. They may hold their babies less, be more likely to let them cry, and, for the most part, be less nurturing. Does this behavior of the mother impact the child later?

Studies, with chimpanzees have demonstrated that anti-social behavior arises from those who are deprived of contact with other chimps. When we see children of less nurturing parents later in life, do they seem more unruly, less socially adapted than other children? While we can speculate that the behavior of these children is due to poor parenting mothers, the child did not receive the right "upbringing," it is just as likely the behavior of the child is due to some epigenetic influence, which may indeed, be a direct result of the mother's low-nurturing behavior.

Can the behavior of the child then be corrected by altering his epigenome? If the child is, in fact, unruly because of the silencing of genes due to methylation brought on by a low-nurturing mother, or some other reason, then would removing the methylation allow for expression of the gene and an altering of behavior? If some behavioral disorders are the result of early environmental conditions, then correcting these conditions could, if not reverse the methylation of genes, at least prevent the methylation of future individuals. Knowing a mother's diet might predispose her children to obesity could give us another tool in understanding how our behavior not only affects us but also those with whom we interact.

In an example of how a mother's behavior can influence her child's development, in what seems to be a very epigenetic way, recent data indicates obesity increases the risk of autism. Dr. Paula Krakowski of the University of California at Davis, investigated nearly 700 children ages 2 to 5 with either autism spectral disorder, or developmental delays. As controls, she looked at another 315 children with normal development.

She found that women with metabolic conditions, diabetes and obesity, while pregnant, had a 67% greater chance of having a child with autism than did women of normal weight. The chance of having an autistic child jumped from 1 in 88 to 1 in 53 for obese women. The mother's blood glucose levels have a direct impact on the fetus; prolonged exposure to elevated levels, (due to the mother's insulin resistance) ultimately leads to a low oxygen condition and an iron deficiency in the fetal tissues. These conditions affect neurodevelopment, myelination and connectivity in the hippocampus region of the brain.[93]

It is easy to imagine that the mother's environment is likely also having an epigenetic influence on the development of the embryo/fetus. While not yet demonstrated, the methylation of a gene may likely contribute to the conditions of autism and this methylation, a direct result of the mother's environment. The mother, by her behavior/actions (in this case being obese) determines, in part, the neurological development of her children, which in turn will have profound impact on the behavior of those children.

In an amazing example of how changing a mother's environment can have a direct impact on her offspring, a recent study has shown that obese women have thinner children after having undergone weight-loss surgery. In looking at women who had had children before and after having had such surgery, the researchers discovered that siblings born after mom's weight reduction were slimmer than their older siblings. The researchers determined that 5,698 genes were differentially methylated between offspring born before and after surgery.[94] That's right! Nearly 5,700 genes saw epigenetic changes between siblings born before and after the mother's weight loss.

Assuming estimates that we have about 25,000 genes this means more than 20% were modified epigenetically by the mother's physiology. Several genes related to weight gain (and other related health issues) appear to have been turned off in the younger siblings. Of course, many factors play a role in one's weight, but obviously having the right genes being turned-on or off can be a huge advantage.

Effects from Grandpa's Generation

In the cases mentioned above—the agouti gene, the low-licking mothers, and the obese mothers—the methylation of the gene was a matter of the environment of the cell, rather than inherited through generations. In the case of the agouti gene, the mother's diet was influencing the molecules to be found in the microenvironment of the cells of the developing fetus. In the case of the low-lickers, the mother's behavior was causing brain chemistry in her offspring that led to methylation of the stress suppressor hormone. And, as we see, obese women are also altering the epigenome of their offspring. So, the mother can cause the methylation of her offspring's genes through her actions, but can such methylation be passed on to future generations?

To determine this you would need health records going back generations. Fortunately, such an archive exists for the small Swedish village, Overkalix. Here they have been keeping track of birth and death rates, including cause, when known, for centuries. Additionally, they have tracked harvests; reporting food supply for each year. Over the centuries, the region has weathered many famines; the populations living there, because of their isolation, have experienced periods of good bounty when harvests were high, but have also faced severe hardship when harvests were low. While studying the records, over a 20 year period, Lars Olov Bygren, a public health expert in Sweden, noticed something peculiar: it seemed that famine might affect people generations later; even if they had never experienced famine.[95]

In collaboration with geneticist Marcus Pembrey, Bygren began looking through the records for deaths caused by diabetes. He then went back to the grandparents to see if there was anything unusual about their diets. What he found was astounding. If harvests were good and the grandfather had plenty to eat, then there was a four times greater chance the grandson would die from diabetes related illness. For the first time, we are seeing the environment of the male affect his offspring.[96]

Pembrey and Bygren were able to look at food supply for each of the first 20 years of the grandparent's life. What they saw was that there seem to be periods in the grandparent's development in which they were susceptible to diet-induced, epigenetic modifications. It seems the grandfather was most susceptible to changes if famine occurred during the years he was experiencing puberty, while the grandmother was susceptible while she was still in the womb.

Interestingly, but maybe not surprisingly, these periods of susceptibility occur during initial sperm formation for the male and egg development for the female. It appears as if environmental conditions during the time of gamete formation can cause the methylation (switching on and off) of genes, which then can be passed trans-generationally. This means the methylation of a gene continues and is conserved during the replication on the DNA. The environment of the grandparents is determining gene expression for the grandchild.

While Lamarck may not have hit on the mechanism, his idea that parents could acquire properties from exposure to environmental conditions and then pass those properties to the following generations, has been validated with the discovery of epigenetics. Again, we see our choices being limited not by anything we do, but by the environment our cells find themselves in, and now, by the environment our grandparents' cells encountered.

PART TWO

EXPERIENCE, DEVELOPMENT AND BEHAVIOR

CHAPTER 8

Detection

"A neuron's activity is a function of its own past activity."

—BRUCE J. MACLENNON

A Collection of Individuals

If you have ever found yourself in a crushing crowd, maybe like the one my daughter and I found ourselves in at the Giants' World Series Championship celebration, you realize, in many ways, you are at the mercy of the crowd. When the crowd sways, you are forced to sway along. Moving against the crowd can almost be impossible. And so, although you are very much your own person, you have also become a member of a much larger collection of individuals. The first part of this book examined how a cell's response is governed by its environmental conditions, but organisms are a collection of cells acting in a coordinated fashion. How these cells interact with each other is what will produce the behavior of the organism. Each cell's response is individual, but it is being greatly influenced by the other cells around it.

Now we will examine how a collection of cells, in response to external stimuli, determines the behavior of the organism. I should point out that all of our behavior is reactive, despite what motivational speakers might say about being proactive, it is us responding to the stimuli we are receiving and have received.

The "Thinking" Cell

Nerve cells are the thinking cells of the body: they allow us to sense our environment and then allow us to respond to our environment. In many ways, nerve cells are similar in function to the cell membrane. The cell membrane receives stimuli from its microenvironment to which the cell responds; nerve cells receive stimuli from the organism's macroenvironment to which the organism responds. We are looking at different levels of environmental response: "micro" in the case of the cell and "macro" in the case of the nervous system. However, we must remember that while the nervous system will function in a coordinated fashion in response to the macroenvironment, it is still composed of individual cells, each responding to their microenvironment.

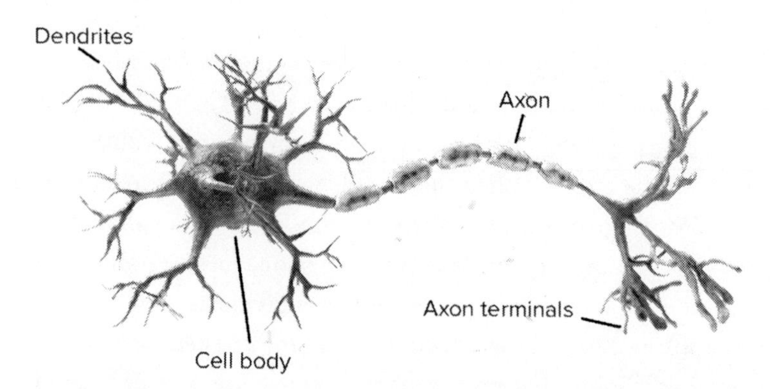

FIGURE 9: "The thinking cell." A standard neuron; with its many dendrites the cell receives stimuli from several other neurons, and with its branching axon terminals relays this information to other cells.

A nerve cell is composed of three main parts: the dendrites, the cell body, and the axon. The dendrites are branched extension of the cell and the structure that will receive the incoming stimuli (message). Cells can have from one to hundreds, even thousands, of dendritic extensions; each dendrite allowing the cell to receive information from other neurons. The nerve cells of the cerebral cortex have many, many short dendritic extensions permitting the ability to receive information from several different sources. It is suggested that increased stimulation of any particular dendrite causes increased size and conduction. Like a muscle, it gets bigger and faster. Such increase is thought to aid in learning and memory. It is the extensive network of dendrites that allows our nervous system to exchange information so vastly among cells.

The signal being received by the dendrites is in the form of a chemical message. The cell turns this message into an electrical current that will travel the length of the cell and cause another chemical release. On the membrane surface of the dendrites are receptors for a class of chemicals known as neurotransmitters. When enough receptor sites are bound by neurotransmitter molecules, the membrane depolarizes: sodium gates open allowing sodium to rush into the cell, changing the electrical charge at that location and initiating an impulse that travels the length of the cell. There is no partial firing of nerve cells: they either fire or they don't. It will take a certain number of bound receptors before the membrane will depolarize and fire; allowing for the generation of conduction of the impulse.

As you might expect, a dendrite with increased stimulation might increase its concentration of surface receptors for a specific neurotransmitter, thus increasing its rate of response and conduction. Those neurons with a greater number of receptors for any specific molecule will be more sensitive to concentration increases in that molecule. For example, cells in the amygdala of the brain have higher numbers of receptors for oxytocin, especially in women,[97] so when oxytocin concentrations are increased, usually through some external stimuli causing other cells to release it, the cells of the amygdala become more active.

The cell body of the neuron houses the nucleus and a majority of the cytoplasm and organelles. Depending on the type of nerve cell, the cell body can be located near the dendritic end, centrally located, or even raised off the axon. It is here, within the cell body, where the mRNA necessary for protein synthesis and the production of neurotransmitters occurs.

The axon of the nerve cell can be thought of as a long cable used to transmit the electrical impulse. The depolarization of the cell begins at the dendritic end and then travels the length of the axon as the impulse moves down the cell. In some cases, the axon can be extraordinarily long. At the end of the axons are multiple small branches, each ending in a little knob. These knobs contain many vesicles, each filled with neurotransmitter molecules. As the electrical impulse reaches the axon knobs it triggers an influx of calcium ions that causes these vesicles to move to and fuse with the membrane (presynaptic membrane). This fusing releases neurotransmitters into the small space between the axon and dendrites of the neighboring cell. This space between cells is called the synaptic cleft and must be traversed by the neurotransmitters to reach the neighboring cell. The neighboring cell will be studded with surface receptors specific for the neurotransmitters being released. When enough receptors are bound by neurotransmitters, the membrane opens up the sodium gates, depolarizing, and causing the conduction of an electrical impulse—thus transmitting the message from one neuron to the next (Figure 10).

It is at the synaptic cleft where many drugs have their effect. If a molecule can bind to a cell's receptors it may be able to cause the cell to fire an impulse. Some drugs mimic neurotransmitters and stimulate excitability. Other drugs can bind to neurotransmitters and prevent them from binding to the surface receptors of the cell, thus preventing the impulse from being transmitted. Just about every combination of enhancement or depression can take place, given the right kind of molecules.

We take drugs because they affect our nerve cell response. Not just illegal drugs, but everyday products as well: coffee, cigarettes,

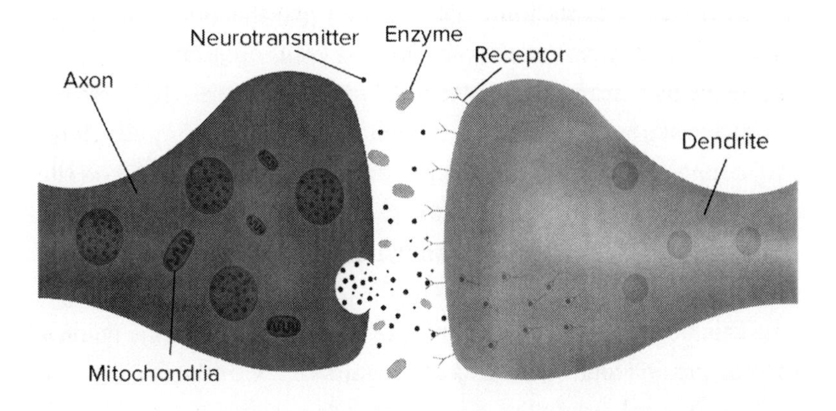

FIGURE 10: Diagram of a synapse; presynaptic axon knob showing vesicle release of neurotransmitters into synaptic space, and neurotransmitters binding to surface receptors of the postsynaptic dendrite.

whiskey. All of these, and any other molecules we are being exposed to: air pollution, medicine, diet, have a direct or indirect effect on our nerve cell's microenvironment. These changes can, and likely will, affect how certain cells respond and ultimately how the organism responds.

Nerve cells can also be triggered by forces other than chemicals. Pressure, as in touch, can stimulate nerve impulses. Electromagnetic waves (light) can cause chemical changes in rods and cones of the eye, therefore triggering an electrical conduction. Sound waves, which apply pressure, can trigger an impulse. In some cases, an electrical impulse can travel between cells by way of gap junctions and ion channels in the cell membranes, but, generally, transmission of the incoming stimuli is by way of chemical messengers.

Because the nerve impulse travels in one direction (this is a general rule with some exceptions), nerve cell function has specialized depending on the direction of impulse. Some cells are sensory neurons; detecting the external environment and relaying this data to the brain. Others are motor neurons, reacting to the incoming information. And many are interneurons, serving as relay stations. These interneurons are concentrated in the brain and spinal cord and are more than just relay stations. They certainly play that role in the spinal cord and are the reason we can pull our hand off a hot stove even before we have realized we have just burned ourselves. They give us our reflex action. However, in the brain, with their many dendrites interconnecting with many other neurons, interneurons take on the role of integrator, or "thinker." It is these cells of the cerebral cortex that are evaluating all the incoming stimuli that tells you later you have just burned your hand.

Thinking is no more than the back and forth firing between neurons in the cerebral cortex and other parts of the brain. Clearly, one single cell is not producing thought, but the electrical and chemical energy exchange between billions of these special interneuron cells does. I am dubious of the notion that thought is anything more than biological. Certainly, we like to believe our thoughts are other worldly, metaphysical, spiritual, non-local (in the case of quantum mechanics), but they are no more than the interplay between interneurons and the chemicals in their environment. Damage these cells or interfere with the neurotransmitter chemistry (or any of the biology for that matter) and you can affect thought; affect thought and you affect behavior.

In studies Paul J. Zak carried out, small amounts of the neurochemical oxytocin added to the nasal passage of subjects caused them to be more trusting and more generous.[98] Zak's studies show it is quite easy to manipulate the behavior of individuals simply by using chemicals that interact with the synapse of nerve cells or the neurotransmitters used for message transmission. And while the behavior may feel completely free to the individual, "I chose to be generous, I chose to be trusting," it is clearly influenced, and manipulated, by changes to the microenvironment of certain nerve cells.

Early Nerve Cells

I am extremely fortunate to live on the Monterey coast in California. This affords me the opportunity to occasionally visit the tide pools with visiting friends and family. As the tide rolls out a wondrous world of intertidal life is revealed. One of the more prominent organisms in most tides pools is the sea anemone. These creatures are found, sometimes in the thousands, attached to the rocks and other substrates by a pedal disc with their many tentacles extending into the water (think of an upside down jellyfish).

It's always fun to have reluctant guests touch these tentacles with their fingers. The first response after doing so is to quickly withdraw the finger from the "sticky" tentacles; the second response is to watch the creature itself retract and pull in those tentacles. I explain to my guest that the stickiness they feel is not from some kind of sugary solution we often associate with being sticky, but from the firing of hundreds of little harpoons, called nematocysts, that actually, momentarily stick into their finger. I go on to describe the reaction of the sea anemone, the retraction of the tentacles, as being the result of the most primitive nerve system in the animal kingdom. However, the process that causes my friends to quickly withdraw their fingers is not that much different from the one that causes the retraction of the sea anemone tentacles.

When we look at evolutionary history, we see Cnidarians (jellyfish and sea anemones) as the earliest group of organisms to have cells that can clearly be identified as nerve cells. These cells are arranged such that they resemble a net (called a nerve net). When stimulated, the cells conduct an electrical impulse that causes the organisms to contract. These primitive nerve cells are non-polar for the most part, sending the electrical impulse in both directions. You'll recall our nerve cells are primarily bipolar, or multipolar, with the impulse sent in one direction. Some Cnidarians do have a few cells that exhibit bipolarity, but predominantly, the cells are non-polar. Transmission of the impulse then is bi-directional, radiating away from the point of stimulus, and across synapse by way of neurotransmitters. So, poking the tentacles with your fingers sends an electrical impulse radiating from the point of contact. We can watch the tentacles on one side of the animal retract and pull in, while the tentacles of the other side remain extended. Cnidarians are the earliest multicellular invertebrates to exhibit a behavior, retraction from a stimulus (which is universal to all cells) that is a result of highly specialized cells for sensing the environment.

Sponges, considered more primitive than Cnidarians, have no cells that qualify as nerve cells. They do, however, possess qualities that may be considered preliminary to the evolution of nerve cells. For instance, the hexactinellid sponges generate a calcium wave that propagates over the outer and inner surfaces of the sponge. This calcium wave could be described as an electrical impulse. Additionally, sponges have many genes coding for proteins that are postsynaptic[99] in organisms that do have nerve cells.[100] While no cell qualifies as a nerve cell, we do see the rudimentary parts of a nerve cell begin to appear in sponges.

Moving up the evolutionary ladder beyond Cnidarians, we come to the flatworms, *Platyhelminthes*. You might remember flatworms from your high school or college introductory biology course. A favorite specimen of these beginning biology courses are the Planarians. These flatworms are fun because of their regenerative ability. You can cut a planarian in half and each half will regenerate into a whole organism. Cut a split down the center of the head, but stop midway through to the midsection, and the organism will grow two heads.

It is in the *Platyhelminthes* that we see for the first time anything resembling a brain. The nervous system of these organisms is arranged in more of a ladder-like fashion with an anterior concentration on nerve cells called the cerebral ganglion. In the more complex flatworms, peripheral nerve cells are arranged in increasing concentration to form longitudinal nerve cords. Thus, we have the development of a central nervous system and a peripheral nervous system. We see several new types of stimulus receptors associated with detecting and responding to the environment: tactile receptors, chemoreceptors, even photoreceptors. The impulse from these receptors is being transmitted unidirectionally and there is a distinct division of labor; some neurons are sensory and some neurons are motor. Really, the fundamental structure of the nervous system for the animal kingdom going forward, evolutionarily, is established with *Platyhelminthes*: cephalization, one-way signaling, sensory and motor response.

In addition to the fundamental structure of the nervous system being established, several researchers over the years have identified a number of chemicals that, while they cannot be classified as neurotransmitters in the classical sense, because we are not sure of their action, they are the same as neurotransmitters we possess. For example, serotonin and dopamine, two of our brain chemicals, have been isolated from nerve cells of Cnidarians[101] and *Platyhelminthes.*[102] Acetylcholine, the neurotransmitter we use for muscle contraction, is also found in jellyfish and flatworms and may play a role as an inhibitory neuromuscular transmitter.

Clearly, these very early, very primitive nervous systems are not that much different than ours. In fact, much of what we know about nerve cells comes from studies of two invertebrates, squid and the sea slug Aplysia. Our nervous system is much more complicated, with modifications and specialization, but fundamentally, the action of a nerve cell, be it in a lowly flatworm or a highly organized human, is the same. These cells receive stimuli from the environment, transmit an impulse to a central ganglion by way of neurotransmitters, and then respond by way of a motor neuron.

After watching Planarians for a while, it's easy to start to anthropomorphize that the organism is moving with purpose and thinking about what it is doing. We can watch it respond to chemicals as it moves toward a food source, or we can watch it react to light, or to touch. But of course instead of thinking, the organism is responding to the cues of its environment. The flatworms' more advanced nervous system allows it to integrate the incoming information from the sensory neurons into its cerebral ganglion and then allows it to send out to the cells, via motor neurons, a coordinated response. How is that any different than what our nervous system does? Fundamentally, it's not.

Incidentally, while in the tide pools playing with the sea anemone, be sure to turn over a few of the smaller rocks to look for flatworms crawling over the underside. They can take on various colors, brown, tan, green, orange, and yellow, and often their bodies will move by undulations. Some will be short and broad, others might be long and skinny; they will all be flat. When finished observing, please be sure to replace the rock.

Senses

Our senses are the means by which we perceive our environment. It is through our senses that we are aware of the world around us. Our eyes receive light stimuli, our ears detect sound waves, both our taste and smell senses are responding to chemical stimuli and our touch sense to pressure and temperature. These senses allow us to observe our immediate surroundings and then respond in a manner appropriate for the situation. This is an ongoing, continuous process. Every moment of consciousness is spent interpreting and responding to the events being sensed.

For us to receive an external stimulus using one of these senses requires the firing of an impulse by the sensory neuron. This means the stimulus must cause the neuron's membrane to depolarize so that an electrical impulse can be transmitted to other neurons. The transmission of the electrical impulse continues on to other neurons until it reaches the brain, where, with other incoming electrical impulses, the information is integrated and a response formulated. Note, it takes time for an impulse to travel from the sense cell to the brain—even if it is just milliseconds; estimates vary depending on the nerve.[103]

This means we are not aware of any event at the moment it happens, but shortly thereafter. It is much like the signal sent to our TV; there is always a delay from when something happens to the time it reaches our television. The signal has to travel across cables and through space to reach our TV. This all takes time. We are aware of the touchdown several moments after it has actually been scored. Imagine what it might mean if your neurons could fire just a few fractions of a millisecond faster than everyone else's; it would be like everyone was moving in slow motion—you would see events unfold before anyone else would.

You might wonder how it is that everything seems to happen instantaneous if indeed it takes all this time for impulses to be transmitted.

The simple fact is the time it takes for interactions of chemicals and cells to occur is irrelevant to our perception. We only become aware of the stimulus at the end of the impulse. Anything happening prior has not been perceived and so we would not be aware. Everything seems to occur instantaneous because awareness can only happen once the transmission process is complete.

The key to our senses is their ability to perceive stimuli. This perception takes place at the cellular level. When light strikes the photoreceptors of the eyes it causes a protein called rhodopsin to change its shape. This change ultimately results in the production of an impulse. Movement of air causes a vibration of our tympanic membrane (eardrum), which moves three tiny bones in the middle ear that moves fluid in the inner ear, changing the position of hair cells there—resulting in an impulse. Small molecules in the air and on the stuff we put in our mouths adhere to surface receptors on olfactory cells and taste buds—resulting in an impulse. The pressure of a touch bends and stretches the cell membrane, increasing permeability to sodium and potassium ions—resulting in an impulse. Everything we perceive is the result of an impulse being generated at the cellular level. If the stimulus is not enough to cause an impulse, then we are not aware of the event. Events we are not aware of do not affect our behavior. I want to emphasize this; we do not respond to events we have not detected, in one way or another, through our senses. Our behavior is being determined by the environment being sensed.

This should put to rest the question of, "If a tree falls in the forest and no one is there to hear it, does it make a sound?" To make a sound would require a device (ears) that can perceive the air movements created when the tree fell, and then translate those air movements into meaningful patterns. Of course, the tree still causes sound waves, but how these waves are sensed is dependent on the receiver. If the stimulus is not sensed (no cells are triggered, no action potential fired, no electrical impulse generated), then the stimulus will not impact the organism's behavior. We do not respond to events we are unaware of.

This is not to say the event does not impact the organism, it may very well do so; it just does not impact the organism's behavior. How can that be? Well, imagine we are in the forest where that tree is about to fall. If, as it is falling, our senses fail to detect it while we jog under (maybe we have our iPod on, earbuds in, and we are rocking out to Seether) and the tree falls and crushes us! Well, the event did not affect the organism's behavior, but certainly affected the organism.

Our behavior is going to be determined by what we perceive, what our senses are able to detect. This detection of the external environment happens at the cellular level with the depolarizing of the membrane of nerve cells. Without this depolarization, the external environment will not impact the organism's behavior, although it may still impact the organism.

Sixth Sense

There is a "sixth sense" I want to mention that, while it is understood by most biologists, is rarely covered when discussing receiving external stimuli. No, it's not metaphysical or spiritual, it's cellular. The cell membrane actually serves as a sixth sense. However, it is not a sense we are necessarily cognitively aware of. The other five senses[104] happen at a level at which, for the most part, we are consciously aware. But the cell membrane may be responding to molecules in the environment of which we are not cognitively aware.

Molecules inhaled or ingested may not stimulate the smell or taste receptors because the cell membrane of those particular cells may lack the appropriate receptors. But these inhaled or ingested molecules may still bind to surface receptors of other cells such as epithelial cells in the lungs or intestines. In this way the organism and its behavior may be affected by molecules detected by cells other than neurons.

This is how some drugs, infections, pollutants can work; they affect cells without the organism necessarily being aware of the stimulus. Viruses can give us the flu, carbon monoxide can cause asphyxiation, aspirin can lead to the relief of a headache; none of these actions requires the direct input of any of our five senses. All are a result of cell membranes interacting with their microenvironments. Of course, once each cell receives and responds to the molecules of its surroundings, other cells will likely do as well, and the behavior of the organism will follow. Flu symptoms can lead the organism to seek bed rest, carbon monoxide poisoning can lead to death, and relief from a headache can improve mood.

In many ways, the cell membrane can be considered the elementary decider.[105] The surface receptors and gated ion channels allow the cell to sense the environment. In the case of neurons, when enough surface receptors are bound, ion channels open and an impulse fires. But the cell is not really making a decision as much as it is responding to the concentration of molecules in the environment. When enough receptors are filled, the neuron fires, regardless—it's not a matter of choice, but an action determined by the cell membrane's interaction with the molecules of the microenvironment.

As another example of the cell membrane's role as decision maker, consider briefly the function of our immune cells. These cells, which are vital to the health of the organism, are continually determining which cells should live and which cells should die based on the molecules that bind to their receptors. Cytotoxic T-cells will crawl over other cells checking their MHC molecules for any peculiarity. If one is found, say a molecule suggesting a virus has invaded or the cell has turned cancerous, then the T-cell will release chemicals which will lead to the death of that cell. Again, the response from the T-cell will only be in regards to whether or not it recognizes a foreign molecule. No "choice" is being made. The response will be determined by the presence or absence of certain molecules. The cell membrane responds to the environment without us necessarily being cognitively aware of such activity.

Perception

Our senses allow us to perceive the world we live in and they define our individual universe. Our universe consists only of what we currently perceive and the experiences we can recall, which can be conscious or unconscious. Can you know something that cannot be perceived? Sure, we can read about wondrous things and make-believe places, but our "vision" of such places is restricted to our experiences. These experiences have been defined by our senses. For most of us, our universe is actually a very small world, limited to what we can and have experienced.

Because our universe is limited to what we perceive and our perceptions based on our senses, each of us will perceive the phenomena around us differently. Sometimes these differences can be very subtle, other times they can be quite extreme. Imagine identical twins standing next to each other looking at a rainbow. Most would agree that they are seeing the same rainbow and therefore having similar experiences.

While it is true they may be having similar experiences, it can be argued that they are actually viewing two very separate rainbows. Because of the way light is refracted when hitting water, the detection of the light spectrum created will be dependent on the positioning of the observer. Twins standing next to each other will actually be seeing different rainbows.

In fact, the only way an experience can be completely shared is if identical observers are sharing the exact space. That is not possible, so each and every experience will be unique to each and every individual. If we wanted to speak of the rainbow objectively, we realize it doesn't really exist. There is no rainbow, per se, only light and water. The rainbow is merely a construct of our interpretation of our perceptions.

So, if identical twins viewing the same rainbow can be seen as having subtle differences in perception, and therefore experiences, then what would be an extreme difference? One can easily imagine the loss of sight or deafness as providing a case for extreme differences in perception and experiences. Can people who have been blind their whole life really imagine the vividness of fire engine red? The subtle difference between teal and turquoise?

We must also recognize our perception of current events will be influenced by past perceived events. No perception happens without recalling other experiences; these experiences help in interpreting and responding to what is currently being perceived.

As an instructor, I am often annoyed by the forgetful students who fail to turn off their cell phones prior to coming into class. It is well known that as we age we lose the ability to hear sound waves of certain frequencies. My daughter and niece, to their great amusement, have demonstrated this many times with ringtones I cannot hear. A discerning student might recognize this phenomenon, set the ring tone on their phone to a frequency I can no longer hear, and never have to worry of annoying me again. The same might not be said for fellow students who may still be interrupted and annoyed by the ringing of the phone. Unaware of the fact I cannot hear the frequency, these students become irritated with me for allowing such interruptions to occur during class. Small changes in our ability to perceive the world around us have significant impact on our experiences.

Our experiences then are what we perceive; these perceptions become our reality and our behavior will adjust accordingly. A not so subtle point to make here is that our perception may not actually reflect reality, and certainly not the perceived reality of others around us. For example, return to the student who has set her phone's ringtone to a frequency I cannot hear. When her phone goes off in class I am unaware of it and so my behavior is unchanged. However, those students that can hear her phone ring may be disturbed by the noise, enough so that they speak to her, or me, after class. Two different realities as a result of what our senses are able to perceive.

This likely raises the question, "Is there one true reality?" I ask, with regards to what and whom? As far as the individual is concerned, her perceptions are real to her and, therefore, every organism's perception is reality. Others' perceptions and concepts can be constructed in our minds upon hearing an account, and coupled with what we've experienced, give us a better record of events, but ultimately our reality is limited to our experiences. So, for the individual, her reality is the true reality. Perceived reality is a matter of individual perception and is subjective.

Is there one universal, objective reality? There are likely levels of reality depending on one's position. Reality is relative. Newtonian physics defines our reality based on the laws that govern how objects interact, suggesting an objective reality. Quantum Mechanics tells us reality is dependent on the observer, suggesting a subjective reality. But, the relevance of these realities to the organism is significant only in how they illicit response. The only thing that is going to influence the organism's behavior is how its cells are reacting to the environment the organism is encountering; both classical and quantum. It is only what is immediately being sensed and perceived, coupled with past experiences of what has been sensed and perceived, that will determine the organism's response.

Let's also recognize that our response, our behavior, is to how things are currently being perceived. Past experiences will influence the response, but it is the current environment that is being perceived. The senses, and the cells making up the senses, are detecting current events. Each event is a new event and each event is triggering the response of a different combination of cells. In no perception can we expect the same stimulus to trigger the exact same combination of cells. Even if the exact same combination of sense receptors is stimulated, the corresponding neurons in the brain would be different. If the event has been experienced before, that alone will cause the firing of neurons that did not fire the first time the event was experienced.

Recall the first time you saw your significant other; you likely had a flood of neurotransmitters based on your attraction to that person and their response to you. This flood of vasopressin, oxytocin, and dopamine lead to an infatuation. Over time, that infatuation grew to what we might describe as love. Love, is a different combination of neurotransmitters, more heavily influenced by oxytocin.[106] You do not have the same reaction each and every time you see your partner as you did that first time. The type of cells being stimulated and the combination of neurotransmitters released changes over time. The experience alone changes the next experience.

This is true of every event. The second time doing something, while sometimes better (think sex here) is never the same as the first one. The first time you taste something, the first time you feel something, the first time you see something, you are establishing cellular changes (memory) that will prevent you from ever experiencing that event exactly the same again.

No two people can ever experience an event exactly the same; the simple fact they are not occupying the same space will cause the firing of a different combination of neurons. They may report similarities in the experience but each experience will also have subtle differences. Additionally, no one person can ever have the same experience twice; they may report similarities in the events, but each will stimulate a different combination of neurons. Each experience will influence the next experience.

The organism's behavior is ultimately being determined by what it senses and perceives. This will be unique for each individual and is the result of the activity of their neurons, which, in turn, are responding to changes in their microenvironment. This cellular activity can cause permanent changes to cell structures as a result of the experience. These permanent changes are reflected in the memories formed. These memories then, as a result of cellular activity based on experience, influence (determine) the organism's behavior going forward.

CHAPTER **9**

Remembering

"Some choices we live not only once but a thousand times over, remembering them for the rest of our lives."

—RICHARD BACH

How Could I Forget?

I consider myself to have a pretty good memory—except for when I do not. As a college student I could easily remember facts and images; usually ten minutes of studying a particular diagram would be enough time for me to memorize and reproduce it. Even today, I can usually have most names of my 80 some students, whom I only see twice a week, memorized by the third week of class. During the semester, I'm pretty good at keeping those names straight with only the occasional flub-up. However, shortly after the semester ends, sometimes within days, I will have forgotten most of the names I spent time learning just three months earlier. I will recognize faces, but the names usually escape me. Even with students I've interacted with several times. The passing of time will severely limit my ability to recall their names. I tell my students at the end of the semester, if you see me around town and want to say hi, please tell me your name; I'll recognize your face, but most assuredly I will have forgotten your name.

Our memories, what we are able to recall at any given moment, will have a significant influence on our behavior. In every situation our reaction to events will rely on the memories we retrieve; sometimes consciously, always unconsciously. However, memories are no more than the interaction between cellular structures and chemicals. What we are able to recall is a matter of which cells are active. Because behavior is in the moment, what we are able to remember at that moment will dictate our actions.

Much of our discussion has focused on the cell's response to its microenvironment, but we must realize the organism's response is to the macroenvironment—and memory is part of that macroenvironment. An EMT responding to an emergency is assessing all that is currently going on around him in addition to relying on the memory of his training. Some of his actions will be deliberate as a result of conscious decisions, but much of what he does will be "automatic" as a result of unconscious activity. It is estimated that only 5% of our brain activity is conscious thought while the other 95% is unconscious activity.[107] This unconscious activity greatly influences our decisions. In either case, memory is the result of the cells that are active at any given time.

Our brain is a dense interconnection of nerve cells and their fibers accompanied by a bunch of support cells called astrocytes. We are born with all the brain cells we are going to have,[108] some 100 billion; it's the main reason our heads are so big, relative to our body, at birth. As we grow, develop, and learn, our brain is not increasing in number of cells, it is however, increasing the interconnections between cells. The nerve cells of the cerebral cortex have extensive interconnections with hundreds of other neurons in the brain. These interconnections are the result of dendritic extensions and spines (a small membranous protrusion from a neuron's dendrite) that allow the individual neuron to receive information from other neurons, nearby and far away. It takes time for these dendritic extensions to become established, facilitated by experience and learning. We start with all our neurons, but we must grow the connections (Figure 11).

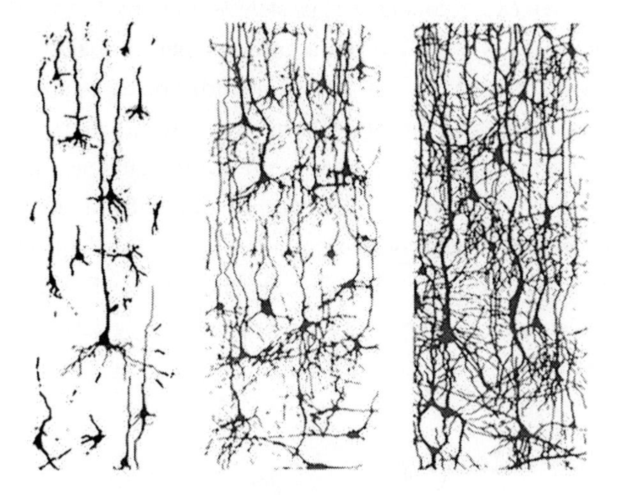

FIGURE 11: Dendrification (increased synaptic connection) at newborn; 6 months; 15 months.

Communication between neurons is facilitated by chemical messengers—neurotransmitters—that travel the short distance across the synapse between two neurons to bind to surface receptors of the target cell. Serotonin has been shown to enhance transmitter release in the presynaptic terminals of sensory neurons.[109] As surface receptors become bound by serotonin, it causes calcium channels to open allowing an influx into the cell. The calcium influx leads to greater transmitter release, which, in turn, leads to the strengthening of synaptic connections between precisely interconnected cells.[110] A synapse is strengthened by an increasing number of surface receptors on the post-synaptic neuron for a particular neurotransmitter. We are growing our connections.

When we remember something, it is because the right combinations of neurons are firing. In fact, studies show that the same neurons used during the original experience are reactivated during memory retrieval. Recall, then, is at the level of individual neurons.[111] Each memory is established by an experience and every experience will conjure memories. These memories will influence how we respond to what we are experiencing and, in fact, become part of what we are experiencing. How we build the memories that ultimately influence our behavior depends on the type of memory being established. We are familiar with, and certainly have experienced, the concepts of short-term and long-term memory; the ideas that you can quickly forget things just learned, or remember things from decades ago. Additionally, other types of memory have been described as working memory (a type of short-term memory used for performing mental math) and emotional memory (long-term memory established somewhat differently).

Short-term Memory

Many times I have experienced forgetting someone's name almost immediately after being introduced. Often, I cannot remember what I had for dinner the previous night. Occasionally, information I know well will be difficult to recall. These are all examples of short-term memory lapses. But, that's what short-term memory is supposed to do, lapse. Short-term memory has been estimated as lasting about 15 seconds[112] and is information we do not retain.

Short-term memory allows us to hold information briefly while we maneuver through our world. These memories consist of two main types of information: sensory memory—what we are perceiving, and long-term memory—what we are considering. In addition to briefly holding what we are currently experiencing, our short-term memory also briefly holds a stream of long-term memories as they are recalled. This can best be thought of as what is happening when we are thinking; we are briefly remembering experiences stored in long-term memory. Short-term memory is fleeting; your brain must quickly evaluate information from both the senses and long-term memories to determine how much of the information is relevant, and needs to be retained. We will see shortly how retention in accomplished, but suffice it to say, we do not hold on to all that we perceive—in fact, on a relative basis, we retain very little of what we experience.

In a classic paper from 1956, George A. Miller demonstrated, on average, people can keep about seven bits of information, plus or minus two, in their short-term memory.[113] Recent studies suggest it may be as low as four bits of information.[114] Either way, short term memories begin to degrade rapidly (unless they are repeated), and are soon replaced by new short-term memories. Your short-term memory is in a constant state of upgrade.

At about the age of twelve we start "chunking" information.[115] Chunking allows us to remember more pieces of information in our short-term memory. By putting groups of information together, we can hold onto four to seven groups each with four to seven bits of information. If you are one that does not make a list, you probably chunk when you go to the grocery store; you mentally store your vegetables together: lettuce, tomatoes, broccoli; your dairy: milk, butter, yogurt; and your necessities: a six pack of Coke and a bottle of Jack Daniels.

In addition to what we are currently sensing, the long-term memories we have established also find their way into our short-term memory—as our thoughts. We are thinking all the time, but our thoughts are transitory. To keep a thought in short term memory requires it be continually "refreshed." I will often find myself, as I am sure we all do, in conversation with a group of friends and as I wait for my turn to speak, forgetting what it is I wanted to say. As the conversation bounces around, I'm thinking, I'll make this comment, then the pause comes to allow for interjection and I will have forgotten the point I wanted to make. The reason, a new thought has entered my short term memory. I will have to refocus on the past conversation to remember the point I wanted to make. I will have to refresh my short-term memory with the thought needed. Several times while writing this book I have had a thought or point I've wanted to make only to forget it when I sat down at the computer. Like the sensory information that is always coming in, so are our thoughts; and like much of the sensory information that is not retained, so it is for most of our thoughts.

Working memory is short-term memory with an emphasis on the manipulation of information instead of passive maintenance. Working memory is used to describe goal-oriented manipulation of information and behavior; when a task is at hand and interruptions are to be minimized. Working memory is more focused and attentive and may serve to direct one's attention to the information that is relevant.[116, 117] Any project you are engaging in relies on your working memory. If you are doing an arts and crafts project, your working memory may know the scissors are to your right, you have three colors of construction paper, the tape is to your . . . "Now, where did I put that tape?"

Many theorists consider short-term memory and working memory as one in the same. Why this information is short lived is thought to be a matter of neurotransmitter activity.[118] As we will see, with long-term memory large structural changes take place within the cell, but no such changes occur in neurons excited during short-term memory. Instead, cells involved in short-term memory release various amounts of neurotransmitters during excitation. It is the depletion of the neurotransmitters available for release that activates short-term memory. The pre-synaptic terminal contains vesicles of neurotransmitters that will be released when calcium influxes into the cell; this happens when the cell membrane has been depolarized. Release of neurotransmitters will continue as long as there are impulses; a large number of stimulations can lead to the release of less neurotransmitter in response to being fired causing a decreased response from the "connected" neuron.[119] When firing ceases, the memory is lost, because communication between specific cells has ceased. This is one of a few competing models for short-term memory; there are still many details to be worked out, but, whichever model wins out, it is clear that short term memory is cellular activity that is stimulus-dependent.

Short-term memory allows us to engage and function in our environment here and now. It is absolutely necessary in assimilating, integrating, and maneuvering through the myriad environmental stimuli. Our behavior is a matter of responding to the information being held in our short-term memory. However, successful behavioral responses also require our short-term memory to receive input from our long-term memory. We are indeed reacting to our immediate environment, briefly holding in our thoughts the events going on around us, but we are also relying on past experiences to help us determine an appropriate response to current events. These past experiences are stored in specific neuronal networks we identify as long-term memory.

Long-term Memory

The establishment of long-term memory, unlike short-term memory, requires structural changes to the neuron. Whereas short-term memory is chemical fluctuations within the cells microenvironment, long-term memory is actual cell construction and requires the formation of new proteins. This synthesis of new proteins serves in strengthening the connections between synapses.[120] Long term storage of memories then is accomplished by expressing the genes that code for the proteins used in strengthening these connections. "Learning results from changes in the strength of the synaptic connections between precisely interconnected cells [and] experience alters the strength and effectiveness of these pre-existing chemical connections."[121]

Researchers, in 2004, demonstrated there was a direct activation signal from the synapse to the synthesis of proteins by way of an enzyme—serving to switch on various genes.[122] These newly synthesized proteins strengthen connections by increasing the number of surface receptors for the specific neurotransmitters at that synapse. Additionally, protein synthesis is occurring on both sides of the synapse; so, gene expression occurs in the cell being excited (post-synaptic) and the cell doing the exciting (pre-synaptic). Protein synthesis in the presynaptic neuron is associated with neurotransmitter production. So, the connection between neurons is strengthened by increased surface receptors on one side of the synapse and increased neurotransmitter release on the other side. Of course, this is a simplification, as the researchers noted, "This process involves 'up-regulating' the synthesis of a large number of proteins."[123] What we are doing is essentially establishing a permanent path to a specific memory. As you might expect, this takes time.

In order for long-term memory to form, our thoughts and perceptions must first pass through short-term memory. This is indeed what happens the first time we are exposed to an idea or stimulus; a neuron in the brain has been excited by a release of neurotransmitters. If not exposed again to that particular stimulus, then neurotransmitter release ceases and uptake begins. The synaptic space is cleared of neurotransmitters, the connection between neurons not strengthened, and the memory forgotten. However, if the stimulus continues, then protein synthesis begins and the connection strengthens.[124]

Although a neuron may have hundreds of dendritic connections, the changes that take place in the cell are synapse specific.[125] Even though long-term memory requires DNA activity, a whole-cell process, the growth in structure and function occurs at the synapse that was stimulated. The binding of serotonin to the receptors of specific synapses serves to mark those synapses such that they can use the protein products resulting from the induced gene expression.[126] If a synapse has not been stimulated by serotonin, it will not be able to use the proteins being made available.

However, a single, simple stimulus is not enough to turn on the genes necessary to strengthen the connection—the stimulus must be repeated. There must be a train of stimuli where one neuron continues to cause the firing of the other neuron. The neuron committed to that memory must be stimulated repeatedly in order for protein synthesis to begin, and ultimately, the strengthening of connections between synapses.

Repetition causes the activation of proteins already present; these then migrate to the nucleus of the cell where they will influence gene expression.[127] But protein synthesis does not begin with the first train of stimuli. Instead, the modification of pre-existing proteins present in the pre-existing synapse strengthens the connection initially.[128] A memory from a single train of stimuli can last anywhere from several minutes to a few hours.[129] This is what happens when you try to remember your schedule of appointments for the day: I need to meet with Bob at 9:00, I have a dentist appointment at 11:00, and I must pick up Susie at 3:00. You likely repeat this schedule to yourself several times in the morning, but unless you revisit this schedule again, by noon you may have forgotten you need to get Susie; especially, if it is not part of a normal routine. The same may be true for a list of names learned at a party; you are able to keep them straight during the evening, but by tomorrow you have forgotten most. This is early phase long-term potentiation (LTP).

On the other hand, late-phase long-term potentiation are memories you may keep for a life time. These memories come from neurons with strong synaptic connections as a result of protein synthesis. In order to establish late phase LTP, the neuron must receive repeated trains of stimuli. These repeated trains of stimuli activate a cascade of reactions leading to transcription and translation of specific genes (protein production).[130] Interestingly, you have to break up your exposure to the stimulus in order for long-term memories to form.[131]

Researchers have shown that space conditioning is necessary for learning. In a study using fruit flies, scientists observed a change in "... calcium influx into a specific set of neurons ..."[132] that was absent in long-term memory impaired flies. This increased calcium influx was identified as a signature component of long-term memory; it served as a memory trace. These memory traces only occurred with space conditioning and paralleled with behavioral changes.[133] Essentially, multiple exposures to a stimulus interspersed with periods of rest greatly improve long-term memory: "This phenomenon of spaced-conditioning is conserved across all species."[134] As a memory is learned and reinforced, it causes the organism to react accordingly: behavior, determined by cellular activity, in response to environmental stimuli.

This really shouldn't be surprising. Most people are aware of how memory works simply through their own experience. I have been studying biology for many years; still, I have to review my notes and reinforce those connections with what I think I know, before every lecture. During my lecture I need notes to help me recall the points I want to make and to keep my presentation structured. I also know if I don't write things down, I will forget them. What must be recognized is that there is a lot of plasticity with regards to our neurons and their connections with other nerve cells is highly dependent on stimulation. There is an "... extensive dialog between synapse and the nucleus, and the nucleus and the synapse."[135] If we are not stimulating the neurons of our long-term memory, those connections, too, can be weakened and lost.

One Memory, One Cell?

Most intriguing, studies show individual neurons can be linked to specific recall. In 1971, John O'Keefe and John Dostrovsky demonstrated that a specific set of cells in the hippocampus served as "place cells."[136] When mice were put into an enclosed space, a specific field of neurons would fire in relation to an internal representation of that space. Each time the mouse was placed in the same environment, the same field of neurons would fire. When placed into a new environment, a new field of neurons established place reference. The same kind of spatial navigation is also observed in the human hippocampus.[137] In fact, all mammals have place cells located in the hippocampus that provide for spatial navigation.

Using fMRI, researchers have observed pattern formation in the recall of spatial memories.[138] Several subjects were monitored while navigating a virtual reality environment. Interestingly, similar patterns of brain activity were observed to correspond with locations within the virtual environment. The researchers knew where the subject was in the maze simply by the pattern of brain activity. These results demonstrated there was a functional structure to the encoding of memories. Individuals that rely heavily on spatial memory, like cabbies, have an enlarged rear area of the hippocampus.[139]

If we know a specific pattern indicates certain thought, could we then be able to read people's minds? Imagine a technology that could sense your brain activity through the electrical waves it generates, quickly translate these patterns into images, and then display these images on a screen. We could make people wear glasses so we could see what someone might be thinking. Could we control our thoughts knowing others could see them? Remember, the thought is generated before you are aware of the thought.

In another example of neuron specificity, it was demonstrated at Mark Mayford's laboratory at the Scripps Research Institute, that the retrieval of a memory activates the same neuron as it did during conditioning. Using fear conditioning on a unique mouse in which researchers can genetically tag individual neurons, they were able to show learning and recall at the level of the individual neuron. This data suggest a stable synaptic change between neurons develops during the initial experience. As Mayford states, we ". . . found neurons in the basolateral amygdala that were reactivated during memory retrieval. The number of reactivated neurons correlated with the behavioral expression of that fear memory [indicating] a stable correlation between these neurons and memory."[140]

More recently, and even more astonishing, researchers have identified individual neurons for very specific recall. Sebastian Seung, in his book *Connectome*,[141] describes a subject in which a single neuron responded only to pictures of Jennifer Aniston. The unique situation of the subject preparing to undergo brain surgery allowed researchers an opportunity to monitor specific neuron activity prior to the operation. What they observed is fascinating and seems perfectly logical. When they showed the subject pictures of celebrities, one specific neuron responded every time to Jennifer Aniston—and only to Jennifer Aniston. The subject was shown pictures of several other celebrities, even a picture of Jennifer Aniston with Brad Pitt, none triggered a response from the "Jennifer Aniston" neuron. Only pictures of Jennifer Aniston caused the excitation of this very specific neuron. The researchers also identified specific neurons for Halle Berry, Julia Roberts, and others.

Is there a celebrity neuron for each celebrity you know? It doesn't seem unreasonable, and evidence seems to suggest indeed that is the case. One can easily imagine a recall system in which individual neurons code for specific images, sounds, taste, etc. Why would we not expect the chain of action potentials to end at a specific nerve cell? Seung suggests that in addition to these chains of neurons, more likely, there are actually assemblies of neurons for a particular thought or recollection. These neuron assemblies, "connectomes," then, are associated with a particular stimulus. It's not that there is one neuron that fires when Jennifer Aniston's image is seen; instead, there is a collection of neurons that fire, which eventually leads to the triggering of the Jennifer Aniston neuron. There will be several different neurons activated by various features of Jennifer (hair, eyes, etc.) that will allow her to be recognized. It's easy to imagine many different assemblies of neurons that might lead to the firing of a Jennifer Aniston neuron: of course an image of her would do this, but her voice might also lead to the firing of this specific neuron, or, depending on how intimate you might be, a scent could cause the firing of a cell assembly leading to this neuron.

This neuron then has been programed to respond when that stimulus is encountered again. There are likely several different pathways to any specific neuron, each causing the recall of a very specific memory, and, likely, also conjuring up various other associated memories. So our subject has a neuron that responds to Jennifer Aniston's photo. Would this same neuron fire if he heard her voice, or if he knew her scent? It is likely there are a number of different stimuli (considering neurons have hundreds of synapses) each with a different pathway, that can excite a particular neuron. At the same time, it seems very reasonable to expect, even assume, there are memory specific, experience specific, brain cells.

Recollection

So, you are sitting there watching Jeopardy when Alex Trabec asks, "Often referred to as 'hitting your funny bone,' there is nothing amusing when you pinch this nerve."[142] At that moment, each of the contestants has running through their working memory the fact they must wait for the answer to be read before they can ring their buzzer, while, simultaneously, each contestant's brain is also sending electrical impulses through various neuronal pathways/assemblies in search for the nerve cell with the right answer. Part of the working memory is also aware that in addition to the answer being read, pressing the buzzer must wait until the question is retrieved.

Contestant three has no idea; she reads and hears the word funny-bone and the first image she gets is of the board game Operation. She thinks, "My brother always cheated when we played that game."

Contestant two thinks he knows the answer and buzzes in first. "What is the humerus?" Before the show, contestant two, an accountant from Idaho, had spent a little time studying the main bones of the body. His recall was still pretty good and when he heard the word funny, his neuronal pathways went to the nerve cell (cell assembly) he had associated with funny when he had read the word humerus earlier. Wrong. Doh!

Contestant one is a nursing student who has just finished an anatomy course. Her first inclination was also humerus, but in hearing it was the wrong response she was able to search various neuron assemblies until the right one came to her. New information, "humerus is the wrong answer" started a new chain of neurons searching for the nerve cell that held the term "ulnar." Given a little more time and knowledge, she was able to find, via the excitation of each nerve cell she considered, the correct response. You can imagine her exciting individual neurons one at a time in her search, first the humerus neuron excites. Nope, wrong answer. Then the neuron for the radius excites. Nope, that's not it. Then the neuron for the ulna excites. A new pathway of neurons is excited by this particular cell being stimulated and you find confirmation with an image you remember from a text book. "What is the ulnar nerve?"

Each contestant's response is solely determined by the cellular activity taking place in the brain. This cellular activity is a result of several factors, including past experiences. These past experiences have wired each contestant such that they can only give the answer they do given the current circumstances. They are not freely choosing their response. All the contestants want to give the right answer, what prevents this from happening is the specific cellular activity of each contestant. Activity has been determined by two things: the genes that were switched on during memory formation, and the environment being experienced at the time—then and now. Each contestant will be limited by their genetic responses to past environments/experiences. Even in reacting to the current situation, you are remembering past events—these memories influence your response.

Emotional Memories

Long-term memory is often divided into declarative memory and procedural memory. Declarative memory has to do with conscious thought, while procedural memory is more concerned with body movement. One type of long-term memory that evokes properties of both is emotional memory. Emotional memory is available consciously and unconsciously (sometimes referred to as subconsciously), and can illicit strong physiological changes in the brain and throughout the body. Emotional memories can be extremely powerful and have great influence over our behavior. The mechanism for establishing emotional memory is similar to any long term memory—protein synthesis. However, the associated physiological changes that take place make the connections significantly stronger, while also increasing and strengthening synaptic connections with other neurons. From our own experience we know emotionally charged events are better remembered.

Emotion plays a significant role in what we remember. Not only do we remember emotion-filled events better, we also remember more detail from these experiences. Pleasant emotions are remembered better than unpleasant ones and our mood at the time of the experience can affect the memory.[143] Surprisingly, the information being remembered isn't as important to memory formation as the emotional state at the time of formation. That is, being happy or sad has a greater influence on memory storage than the significance of the information being stored. Furthermore, it is easier to remember events when your mood matches the time of formation. Pleasant memories are easier to recall when you are happy and unpleasant memories are easier to recall when you are depressed.

Moreover, there are gender and age differences in processing emotional memory. Women are better at remembering emotional memories, which seems to be related to their ability to more tightly integrate emotional significance with the memory.[144] Age also impacts our response to emotional events. The older we get, the less activated is our amygdala by negative images.[145] It could be that as our interconnection of neuronal synapses matures, our cerebral cortex is better able to provide an overriding influence on our amygdala, thus establishing weaker connections with memories of negative feelings.

Emotional memories are considered a type of LTP, however, formation is different than long-term memories associated with studying/learning. Certainly we can have strong recall for events that we only experienced once. Like long-term memory for learning, emotional memory involves the need for protein synthesis to establish new and stronger connections between synapses. Unlike memory of learning, emotional memory involves other systems and input.

The hippocampus is a banana-shaped structure located in the medial temporal lobe on either side of the brain. It is a densely packed layer of neurons with extensive connections throughout the neocortex[146] and other parts of the brain. The hippocampus has been identified as the region of the brain in which short-term memory is "encoded" into long-term memory and is primarily responsible for generating episodic memories. Episodic memories are of specific events and are generally autobiographical: memory of time, place, emotion and contextual knowledge. "Newly acquired memories are gradually transferred to neocortical stores through the process of memory consolidation."[147]

Memory consolidation occurs when synapses strengthen both initially and over time.

In addition to sending memories to the neocortex, the hippocampus also receives input from the neocortex which allows it to quickly search for memories already formed. Nearly all functions performed by the hippocampus are in collaboration with other parts of the brain. One of those parts is the amygdala. (Figure 12)

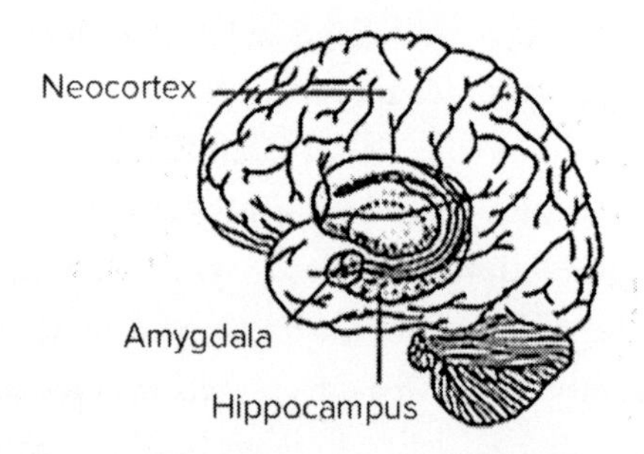

FIGURE 12: This image of the brain shows the location of both the hippocampus and amygdala; each integral in memory formation. the neocortex is where higher level "thinking" occurs.

The amygdala plays a central role in emotion and directly influences the hippocampus during emotionally-charged events. Although not completely understood, it is likely the hormonal release caused by emotional stimuli triggers the amygdala's impact on the hippocampus. The amygdala is responsive for a number of the neurotransmitters released during emotional events, which, in turn, causes the stimulation of the hippocampus. This excitation can lead to long-term potentiation. In addition, emotional events cause the release of hormones that will stay in the system for a period of time. Recall anytime you have been excited or scared, you do not return to a calm state immediately after the stimulus is gone. Instead, one may stay amped up for some time after. During this time, our amygdala can continue to excite the hippocampus; in doing so, the memory becomes more strongly encoded.

Memory is enhanced by the release of hormones during stress, specifically cortisol and epinephrine, two major players in the flight-or-fight response. These hormones also directly influence the amygdala by regulating the release of norepinephrine. The strength of the memory is enhanced by the significance of the experience and is regulated by the release of these stress hormones.[148] People have greater recall of emotionally arousing events. However, stress can also have negative effects on memory. Negative emotional stress can lead to false memories and extended doses of corticoids can damage the hippocampus. So, emotional influence can provide for greater recall within limits, but too much can be harmful.

A couple of other things are notable about the hippocampus; there seems to be structural and functional differences between the sexes. Women have a slightly larger hippocampus, by percentage, than men. Additionally, men use their hippocampus for navigation whereas women use the prefrontal cortex for navigation. There are also developmental changes that take place with age. Studies show that older male adolescents have significantly more gray matter in the left medial lobe and a larger hippocampus than do younger males. This difference implies that our memory improves during late adolescence, as our ability to consolidate information is enhanced by increased synaptic connections.

In an interesting twist, researchers have found a gene that, in those who have it, seems to give better memory.[149] The KIBRA gene comes in two varieties, C and T. Those individuals with at least one KIBRA-T gene perform better on memory tests than those who do not have the gene. It was observed that hippocampus activation was higher in T-carriers, suggesting the gene's importance in memory performance. An inheritable trait for memory formation, determining the individual's ability to remember certain events, ultimately impacts the behavior of the individual.

I have superficially covered the extremely complex, and, as yet, barely understood mechanism of memory. We are still learning much about how it is formed, retrieved, and lost, but everything we do know revolves around cellular activity. Our behavior, how we respond to the situations we are in, depends on what we remember. Our memories result from cellular activity involving DNA transcription as a result of environmental stimuli. If the cellular processes are interfered with either physically, chemically, or developmentally, memory is impaired. Memory greatly affects our behavior and if our memory is impaired than so will be our behavior. If you fail to remember a certain action causes pain, or where to find food, or this guy swindled you, then you will fail to respond appropriately in the future. There is no behavior that isn't influenced by memory. Even people with amnesia are having their behavior influenced by the very lack of a memory.

CHAPTER 10

Emotions

"One can be the master of what one does,
but never of what one feels."

—GUSTAVE FLAUBERT

Emotional Influence

We are constantly making decisions. Many are mundane or automatic—getting out of bed, driving to the store, brushing our teeth, but enough are of significance that we actually contemplate desire and outcome—considering alternatives before choosing. We give thought into the purchase of a new car, a new home, even what to have for dinner, considering price, features and myriad other factors. We like to think that at the end of our thought process our decision is going to be rational. We will have weighed all of the pertinent data and made the right, logical choice.

In terms of cells, what "rational" here means is that our action is being influenced by greater activity from the cortex, which is a matter of increased synaptic connections. The more connections we have—the more we know—the more rational our actions. These connections increase with learning, which comes from studying information on the topic of interest. Any information about the area of interest, be it from a friend, an article, or a TV commercial will weigh on the decision. However, as it turns out, we can't actually make decisions based on rational thought alone; we have to have a small, or sometimes a big, emotional nudge before we can say yes to that shiny new car, or decide what to have for dinner.

We all have an idea of what emotion is. It's the happiness we feel when that special someone likes us or the sadness when they don't, or the jealousy when they like someone else. We may describe some people as being emotional or events as emotional. In either case, emotion implies a reaction, a behavior. When we say emotion, we mean feelings, feelings of happiness or sadness or jealousy—to name a few. The definition of emotion can vary slightly depending on whom one talks to, but generally, emotion is an affective state of consciousness, involving a change in physiological states, brought about by environmental cues. In other words, it is a thought process—happy, sad, jealous—caused by some external event—getting a date, getting dumped, hearing of another—that results in a chemical change within the body—release of hormones, increased heart rate, higher blood pressure. Emotions are fundamental to our behavior; they are the result of ongoing and continuous environmental influences. You could argue that we are in a constant state of emotional flux.

Usually, however, when it appears someone is primarily responding to their feelings, (which, except in the case of brain abnormalities, I don't think is possible) we will suggest they are not being rational. (So inferior are emotions that Spock and the Vulcans built a society on only using rational thought, absent any subjectivity that might be caused by feelings.) Feelings make us do things we might not do otherwise, as if they have some special power over our objectivity, or that expressing them inappropriately is a sign of immaturity or weakness. Emotions are something to control. This idea comes from the notion that emotions are somehow more primitive than rational thought, and while useful in mating and fighting and fleeing, they are not necessary for sound decision making.

Emotions can be considered primitive because activities such as mating, fighting, and fleeing are seen as necessary behavioral responses for survival of nearly all living things. Emotions associated with these responses—happiness, anger, fear—are universal among humans. But, at what level do we see emotion in other animals? Here, scientists are not in agreement. Behaviorists will generally say that a non-human animal's action is conditioned, simply a stimulus-response mechanism. Others argue that many animals, apes, dogs, birds, exhibit behaviors that can only be described as emotion. It is difficult to know without knowing what the animal is thinking. It is easy to look at a dog and see what appears to be emotion. My dog shows excitement when I suggest we go play ball, he cowers when I raise my voice, and (reportedly) seems to show anxiety when I am gone.

There are many stories of grief stricken dogs refusing to leave the casket or grave of their deceased owner. On August 6th 2011, Navy Seal Jon Tumilson was killed in a helicopter crash. His dog refused to leave the side of the casket. Mourners said the dog seem to be grieving. In January 2011, Laeo, a mixed-breed dog became the face of the tragic flooding and landslide of Brazil when she was photographed sitting patiently by her owner's gravesite.

By all accounts, these actions suggest these animals are behaving emotionally. However, my buddy's pet tortoise doesn't do any of these things. When I roll him the ball, he doesn't pounce on it in excitement (at least he doesn't look excited, maybe in his mind he is doing cartwheels), there is no cowering when I yell for him to "get off the patio!" (Not that I actually do this to a turtle regularly, but I did one time to prove a point.) He doesn't respond joyously to my return from a long term absence. We would all agree that turtles don't seem to have emotion, or at least much less than mammals. So, although we may consider emotions primitive, they are not that primitive.

Emotions, at least in mammals, seem to have arisen from the expansion of our olfactory sense.[150] Many early mammals were nocturnal, an adaptive behavior that allowed them to avoid the big lizards during the day. Being nocturnal, mammals experienced an increase in the reliance of olfactory sense to navigate the environment; smell replaced vision as the dominant sense. Overtime the olfactory lobes of the mammalian brain formed pathways that would become part of the limbic system. The limbic system is a set of brain structures that are involved in behavior, emotion, long term memory, and smell. It has long been known that smell and memory are closely associated and odors can often bring strong emotional responses. Our emotions seem to have evolved in conjunction with our increasing use of olfaction.

What about birds? They did not experience the same nocturnal history mammals have and do not have the same adaptive sense of smell. If indeed, birds do exhibit some form of emotion, it seems to be associated with birds that are more social. Many species of birds have intricate social behaviors with some forming long-term, monogamous pair-bonds—mallard ducks and zebra-finches for instance. In cases where emotion is applied to birds, it is almost always in reference to the more socialized birds, and usually in regards to their socio-sexual behavior. If some of these behaviors qualify as emotions, it is likely a result of convergent evolution as a result of the common selection pressures applied by the demands of social interaction.

The evolution of emotion likely coincided with the evolution of sociality. The more social the species, the more expressive they seem to be emotionally. Cats, for the most part, are solitary animals, living individually and coming together primarily to mate. Some cats, like lions, form prides, but that is more the exception. Dogs, on the other hand, seem to be pack oriented. Again, this is a generalization, but accurate in many regards. I think most people would agree dogs are the more emotionally expressive of the two. Dogs are excited to play, excited when you return, and warn of the approach of a stranger. Cats, generally, don't care to be bothered. The more intricate the social interaction between individuals of a species the more varied the emotional behavior seems to be.

Humans have an extremely wide range of emotional conditions, likely a result of the complex nature in which we interact with each other. Our intimate living arrangements require us to interpret and respond to the emotional states of those around us, as well as our own. Emotion serves to facilitate the interactions between the individuals of close social groups. If we are not interacting with other individuals intimately, then there would be little need for emotion on the level experienced by humans.

The Amygdala—The Center of Emotion

Emotion has long been considered an element of the limbic system, which includes, among other structures, the hippocampus, septal nuclei, dentate and cingulate gyrus, and the amygdala. However, studies now point to the amygdala as having a more significant role in processing emotion. In fact, the amygdala is proving to be so integral to our everyday behavior that we must take time here to examine it further. Rhawn Joseph, PhD., an authority on the amygdala, describes it this way, "In contrast to the primitive hypothalamus, the more recently developed amygdala is preeminent in control and mediation of all higher order emotional and motivational activities."[151] Isn't that essentially the essence of our behavior—emotional and motivational activities?

The amygdala is located deep in the temporal lobes of the brain (Figure 12) and has a rich interconnection of neurons with various cortical regions. Part of this rich interconnection of neurons extends to the hypothalamus, thus giving the amygdala direct influence on this structure. The amygdala "is able to modulate hypothalamic activity through inhibitory and excitatory projections to this structure."[152]

The hypothalamus acts directly on the pituitary gland to control the release of many of the hormones responsible for regulating our internal environment. The hypothalamus controls the release of our sex hormones and growth hormones and adrenaline hormones and most of the hormones released by our endocrine system. In addition, the hypothalamus produces oxytocin, a hormone responsible for uterine contraction and milk release, but has also been implicated in pair bonding and what can be considered the emotion of love. Of course, everyone knows hormones make us moody and emotional. It has also been shown that oxytocin plays a significant role in trust and generosity[153]—two qualities vital to the functional social community.

The amygdala then, with its direct control of the hypothalamus, influences behavior. Other animals have a hypothalamus and many of these same hormones, but we would not necessarily say they have emotion. Much of the influence of the hypothalamus, endocrine glands and the hormones they release, act on the autonomic nervous system helping maintain homeostasis; hardly activities that require emotion.

The connection with the hypothalamus gives the amygdala both a reading of the internal environment, as well as a way to modulate it. The amygdala has connections to excite or inhibit the hypothalamus, but the hypothalamus can also stimulate the amygdala. The amygdala and hypothalamus interact with regards to hunger, fear, anger, sexual activity, etc., to meet the needs of the organism. The amygdala, on input from the hypothalamus, will scan the environment in search of what is needed, food, water, a mate, at which time it will direct the hypothalamus to respond accordingly.

The limbic system evolved out of an expansion of the olfactory lobe and the amygdala receives significant input from the olfactory bulbs. Additionally, the amygdala receives input from our taste receptors so that smell and taste sensations come together here. This is likely why odor and food can have such an emotional impact.

Furthermore, the amygdala receives input from tactile sensations and can be excited by the slightest touch. This is why a little touch here or there can be so powerful and generally elicits what is best considered an emotional response. And often, this response is subconscious, with the individual not even aware of the touch or its effect. Smart waitresses know this. It has been shown that those waitresses who touch their customers on the shoulder or arm tend to make more in tips; they are perceived as being more positive.[154] A time tested means of flirting is to lightly touch the person you are interested in. Women can be quite skilled at this, touching a guy on the hand or his forearm to signal interest; and men can be quite responsive. But the reverse scenario is also true. Nicolas Gueguen demonstrated that women were nearly twice as likely to give their phone number to a strange man if he had lightly touched her arm during the request. When interviewed after the encounter, those women who had been touched, said the man seemed more attractive and dominant[155]—qualities women like in men.

Touching of other people, or being touched by others, certainly has an influence on our perceptions, but it does not end there. Apparently, the touching of inanimate objects can also affect the way we think and behave. In two separate studies, the touching or holding of an object influenced the opinion people had of others. In one study,

John Bargh demonstrated that holding a warm cup of coffee versus a cup of iced tea for someone conveyed a sense of warmth about that person.[156] In a related study, Jim Ackerman showed that the weight of a clipboard influenced the opinion of an applicant by those evaluating the resume. Resumes presented on a heavy clipboard were viewed as more qualified than those presented on a light clipboard.[157] In both of these studies, the object influenced the opinion of the participate, even though in reality the object had no bearing on that person's personality or qualifications. Experiencing a touch influences perception of mind.

All this touching contributes to a release of oxytocin. Paul Zak, author of *The Morale Molecule; The Source of Love and Prosperity*, describes how being hugged causes a momentary spike in oxytocin. Many of the cells in areas of the brain associated with emotion, including and especially the amygdala, have greater concentrations of oxytocin receptors. Being hugged or touched lightly imparts a sense of trust, which coincides with oxytocin release from this brief contact, and in many cases made people more generous. Behavior was changed because of cellular changes due to behavior of others. A hug, warmly received, causes oxytocin release from some cells, which then binds to other cells in the emotion circuit, leading to a more trusting behavior.

The senses of smell, taste, and touch feed directly to the emotional centers of our brain as the amygdala uses this input to scan the present environmental conditions. Our response to these stimuli is generally immediate, but can be very subtle. Simply improving your feelings towards a waitress is an example of how subtle this response can be. This is an emotional response, void of any real reason or logic, as a result of environmental cues detected by the amygdala.

Processing the Incoming

The amygdala receives a continuous flow of data from our senses and also has a reciprocal connection with the auditory area of the brain and rich interconnections with the frontal and temporal regions of the brain.[158] These connections allow the amygdala to receive input on what is being seen and heard and then scrutinize the information. The interconnections with the frontal and temporal lobes also allow the amygdala to receive integrated information from the cortex. These are fully formed perceptions of what is being sensed.[159] For instance, this very pretty girl just touched your arm, a tiger a hundred yards away looks hungry. Receiving this information, your amygdala will signal for you to take note of a possible mating chance—the pretty girl, or in the case of the tiger—run!

The amygdala monitors our external environment for emotional and motivational significance and then through its connections with the hypothalamus, the brain stem, and the cortex, influences our internal environment, thus resulting in a behavior or action. Our amygdala is integral to our ability to interact socially and maintain a social network, as well as playing a significant role in our fear and fight or flight response.

Besides influencing, regulating, and coordinating our emotional and motivational responses, the amygdala itself is influenced, partly by the hypothalamus, but also by a host of neurotransmitters. Neurotransmitters such as dopamine, serotonin, acetylcholine, and norepinephrine influence both excitatory and inhibitory neurons of the amygdala. In this way, some control can be exerted on the activity of the amygdala so that we do not respond entirely on emotion. Additionally, various amygdala nuclei have surface receptors for many other types of hormones and neurotransmitters, including glucocorticoid, estrogen, oxytocin, vasopressin, and corticotropin releasing factor.[160, 161] The amygdala also is why drug use may be pleasurable. It has the greatest concentration of opiate receptors. These are just a few of the many receptors associated with neurons of the amygdala. There will likely be others identified in the near future.

If we consider the role of some of these hormones and neurotransmitters, it's easy to see the relative importance the amygdala plays in our behavior. Norepinephrine is a stress hormone and fundamental in our fight or flight response. Dopamine plays a major role in reward-driven learning. Serotonin is involved in mood, hunger and sleep. Oxytocin and vasopressin have been implicated, among other things, as the love hormones. Given how these chemicals will greatly influence which neurons fire, it's easy to imagine how certain levels or combinations of neurotransmitters can determine one's emotion.

The receptors for the above mentioned molecules are unevenly distributed throughout the neurons of the amygdala. The right combination of vasopressin, dopamine and norepinephrine, in concert with the cells having the right receptors, may stimulate one to take a chance and ask that pretty girl out. The wrong combination might cause one to be stressed and tired and not sexually motivated. The level and combination of these chemicals in the microenvironment of the amygdala nuclei will have an overwhelming influence on our behavior.

The Amygdala and the Sexes

The amygdala plays a significant role in the varying sexual behavior seen between men and women. Studies show men and women differ substantially with respect to emotional memory and their responses to sexual stimuli.[162] Males have a larger amygdala than females, an increased number of synaptic connections and a higher concentration of surface receptors for sex hormones.

Adding to the sexual differentiation, females have smaller, more numerous and more densely packed amygdaloid nuclei. Smaller, more densely packed neurons fire more readily.[163]

Other differences include the interconnections between the amygdala and temporal lobes in women and the amygdala and the hypothalamus in men. In females, the connections between the left and right amygdala/temporal lobe are much larger than in males, allowing for greater communication between the two structures. In males, the connections with the hypothalamus are larger, which, not surprisingly, corresponds with the increased number of hormone receptors found in the male amygdala.[164] Furthermore, emotional memories for females were encoded in the left amygdala and for males in the right amygdala.[165]

How are these differences manifested in our behavior? Females are more easily frightened and more emotional (expressively) than males. They can recall emotional memories more quickly, abundantly, vividly and intensely than men, and are more socially sensitive, perceptive, and expressive.[166] These characteristics, which are likely the result of the more easily and more frequently firing neurons, lead women to be more compassionate and maternal.

On the other hand, the amygdala of men shows much more activity to visual sexual stimuli than does the amygdala of women.[167] Men are more responsive to sexually explicit images and display an appetitive motivation to seek sexual reward.[168] Anyone that has watched men behave around pretty women has observed the amygdala in action. Men respond to the attractiveness of a female with the desire to mate her. They are motivated by the chase with sex as the reward. Interestingly, during orgasm, the activity in the left amygdala decreases in men. Women, however, are not motivated by sexually explicit material and their amygdala does not show increase activity when viewing arousing material, nor does it show decreased activity at climax.[169] Men seem to be more driven by the desire of the sexual reward whereas women are more responsive to the obtaining of the reward. Although this dichotomy wreaks havoc with relationships, it's no surprise our brains respond as they do when one considers the reproductive strategies employed for making egg or sperm. Evolutionarily, it would be adaptive for the male to react quickly to visual cues, such as a receptive female, to increase the likelihood of passing on his genes; however, for her, not so much. The female needs to be more selective, more discerning; sexy images are not enough—her reward is the obtaining.[170]

The implications of these sexually motivational differences between men and women are played out every day in our attempts to build relationships. Men, in general, are motivated to have sex, but it's the chase that's the reward. Once the desire is obtained, the motivation subsides (decreased activity in the left amygdala). The focus may even move to a new desire. While for women the reward is consummating, the obtaining—copulation. Women are much less motivated to chase men and much more receptive to being pursued. For women, after climax, the left amygdala activity is steady, her emotional needs are being met. He, on the other hand, has been satiated, and now the amygdala is saying "let's go to sleep."

It is worth noting that the amygdala may also play a role is sexual orientation. It has been observed that homosexual men exhibit more female amygdala patterns than do heterosexual males and homosexual women show more male features than do heterosexual females. The amygdala connections are more widespread from the left amygdala in homosexual males and in heterosexual females; and the connections are more widespread from the right amygdala in homosexual females and heterosexual males.[171] These structural differences suggest an anatomical and physiological influence to homosexuality.

Current evidence seems to indicate some sexual differentiation of the brain occurs during fetal and neonatal development. It is possible that in addition to genetic, there are likely epigenetic influences on sexual orientation. It has been demonstrated that fraternal birth order can have an effect on the sexual orientation of boys.[172] For each additional brother that precedes him, a boy's chances of being gay increases by a third. We have to assume the mother's environment changes with each child. It is likely these changes in her blood chemistry are having an impact on the child's development.

Are chemicals in the microenvironment having an epigenetic effect on the boy's genes? Is there an adaptive benefit to the community or society to reduce the amount of "competitive" males? There is certainly no direct benefit to the individual for being gay. His genes will not be passed on. But maybe there is an indirect benefit. Obviously, he shares genes with others in the group, the more nurturing nature that might come with the developmental difference of the amygdala may aid in the survival of his older siblings. In this way, genes he may share with his brothers will be passed to the next generation.

By all accounts, homosexuality, and heterosexuality for that matter, is not a choice. Ask yourself, who am I attracted to? Do you really have a choice in answering that question? You can't choose to be attracted to someone. Your amygdala responds to the visual stimuli coming in and your behavior adjusts accordingly. We see old, married men do this all the time; a young, pretty lady walks into the room and one may suddenly suck in his stomach, stand a little taller. Why? He's married and has no chance of mating her. Still, the amygdala is firing away with activity so the behavior follows.

Mary's Amygdala

Speaking of behavior and the amygdala, let's revisit Mary, our friend who has recently gone off the pill in order to conceive and now finds herself sexually attracted to her co-worker. We know the amygdala greatly influences emotion and that it has very close ties with the olfactory sense. We also know the amygdala plays a role in our sexual motivation and is influenced by hormones, including estrogen. So, Mary was using birth control pills when she and her husband met and her amygdala was being influenced by the hormones she was taking. This interference confused her amygdala's sexual response. To the amygdala, she was already pregnant, so her sexual needs had changed to someone more nurturing. This was partly determined by the MHC molecules she was detecting. Her amygdala responded favorably to her now husband because his MHC molecules were more preferred given her current internal state, pregnancy.

Women can detect our MHC molecules and favor more similar MHC molecules when pregnant and more dissimilar molecules when not pregnant. Is this what Mary is experiencing? Has going off the pill made her more receptive to her co-workers MHC molecules? Her amygdala is no longer being artificially influenced by the pill. She loves her husband, her amygdala has played a role in establishing some strong emotional memories, but now, this urge to bed "Mr. Dissimilar MHC" is very strong. Her amygdala is firing away every time it senses his MHCs.

Once again, I realize this may be an extreme over simplification for something very complex—relationships. However, I point again to the very circumstantial evidence of pill use and divorce rates near the end of the 60s. While this is only one variable of many that influence both mating and divorce, it does underscore the relationship between cellular activity and behavior. Knowing she is under the influence of her amygdala's excited response to dissimilar MHC molecules, Mary's behavior may change accordingly, avoiding her co-worker, for instance. New knowledge about MHC molecules changes Mary's cellular activity and her neuronal network. Now, when Mr. Dissimilar MHC is nearby, she will understand her attraction to him and her behavior will be altered. Knowledge changes our cellular activity and therefore, our behavior.

CHAPTER 11

Decisions

Reason is "... the slave of the passions."

—DAVID HUME

Reason

Our decisions are made as a single unit, but they are made at the level of trillions of individual cells. It is the collective interaction of many neurons and chemicals that will give rise to the decisions we make. And while many factors will go into any decision—what is happening at that moment, stored memories of similar events, possible repercussions—the action taken is the only one that could have been taken. Our decision is a result of the cellular interpretation of perceived stimuli. The interpretation is collective (trillions of cells acting in coordination) and it is the only one we can reach. Our response is the only response available to us given the situation at the time and our experiences to that time. Every action we take is the manifestation of the collective interaction of our cells with their microenvironment; that microenvironment being influenced by the events of the macroenvironment, now and in the past.

Imagine any major decision you have made in your life—buying your first car, getting married, deciding to have children—at the time of that decision you were weighing all kinds of options and possibilities, looking at the pros and cons, evaluating incoming data with respect to past experiences; in the end, your decision was an amalgamation of all you knew. The decision you made was the only one you could have made. That 1968 Ford Mustang, with the faded yellow paint and the black vinyl top, though it might not have been the smartest purchase, it was the decision my interacting brain cells had settled on and it was the only decision I could make at the time. Now, if I had continued to shop (which also would have been the only decision I could have made if it had been what I had done) as I came across other cars and prices, the information and data changes, and then, certainly, a different decision could have been made. But, at the time of that decision, and any decision, it is the only one our cellular biology will allow.

Consider for a moment reality TV. Now, I'm not much of a fan, but there are a few shows I find entertaining; *Survivor*, for instance. Many of the shows, *Big Brother*, *Jersey Shores*, etc. show people behaving outrageously. Their behaviors are extreme and it is easy to think, "They are faking their reactions, because it is too overblown and nobody behaves that way." I would agree with you that their behavior does not reflect how people normally react, or even how that particular individual would normally react, but I disagree that it is fake. I do not accept behavior can be faked. The behavior is overblown, but the individual is only acting as he or she can, given the situation.

The situation of course is the TV cameras. With the cameras present, the behavior of these fame-seeking, attention-hungry individuals is distorted from how they might have behaved without the cameras. However, it is the only way they can behave with the camera present. Reality TV is not fake, the behavior we see is real; however, it is influenced by the presence of the camera. The people are acting and reacting based on the environment around them, which includes other people doing the same, in response to the camera.

If we think about this for a moment, it means all behavior is real. Exactly! I am not saying people cannot lie, or act, or pretend to be what they are not. What I am saying is the act you are witnessing is a result of the environment the individual is perceiving, which includes many factors: other individuals present, drugs the person might be on, the presence of a camera. At any particular instance of time we only *know* one way to act and that's *how* we act.

All of us have been in situations where our behavior changes because of the circumstances. For example, salty language around the guys in the past would end when ladies come in the room. The change in behavior is a result of the change in the circumstances, which causes different cellular activity, which leads to modified language. Whatever we do, it is being determined by how our cells interpret the changes taking place in their microenvironment.

Emotions and Decision Making

We cannot make purely rational decisions without some sort of emotional input. Studies show that damage to the amygdala, or to the connections the amygdala has with the cortex, can severely inhibit one's ability to make decisions.[173, 174] We often respond emotionally to events before we have had time to think about them. This is the amygdala reacting quickly to the events and environment of the moment. The response is built out of earlier emotional experiences and so the logic behind the response can result from crude associations. Given time to think, your cortex can refine your response, soothe your emotions, or rationalize a more prudent reaction.[175] However, the initial response is often the right one. Neuroscientist, Antonio Damasio, describes the amygdala as being able to cause emotional dumbness by being overreactionary, but also providing emotional wisdom in that its quick response is usually the right one.[176, 177]

I often tell my students, when taking an exam their first inclination is usually the right answer. People often talk of the importance of first impressions; if you consider the first time you meet someone, your initial gut feeling is usually pretty close to how you end up feeling about that person. Studies have shown our opinion of people is generally formed within seconds of initially meeting someone. Nalini Ambady and Robert Rosenthal, in a 1993 study, demonstrated "thin slicing," making quick judgments based on small pieces of information. In their study, individuals were shown ten second video clips of college professors with the sound off. The participates were then asked to rate each professor on 15 qualities: attentiveness, competency, likability, etc. Surprisingly, or maybe not, the ratings the instructors received from these ten second non-verbal snippets of their behavior, closely matched evaluations given at the end of the semester by students who had just completed the full course from the instructor.[178]

Further, evaluations did not change when the video segments were reduced to two seconds. Opinions established early in an encounter seem to be retained. Such quick assessments, for the most part, are emotional judgments. There is little time when making rash decisions like this for rational thought. We are responding to gut feelings. Often these gut feelings are right. There is some character or mannerism that one quickly evaluates; maybe it's their attire, or the way they walk, or their smile. One, or more likely several, of these characteristics is rapidly processed against similar experiences to reach a prejudice within seconds of seeing someone. Our amygdala is influencing these gut reactions, but it is doing so based both on emotional memory and cortex input from previous experiences.

So, our gut feelings have validity; but what about our rational decisions? How does the amygdala influence the decisions we think about?

In 1982, Damasio met Elliot, a patient who had recently had a small tumor removed from his brain, near the frontal lobe. After the surgery, Elliot had difficulty making decisions. He was still just as smart as he was before the surgery, scoring in the 97th percentile on intelligence tests, but he had lost the ability to decide, especially decisions regarding personal and social matters. In incidents described by Damasio, when Elliot was given a choice of where to have dinner, or two dates and times to schedule a meeting, he would have to contemplate all of the significant factors in making his decision: what kind of food did he want, how much did he want to pay, how long would he have to wait, etc. He even went as far as to drive by both restaurants to see how busy they were. In choosing a date to schedule a meeting, Elliot would have to consider all options; doing cost-benefit analysis almost endlessly. Elliot was experiencing "paralysis by analysis." He could not decide, because there was always something left to analyze.[179]

In addition to his inability to make decisions, after the surgery, Elliot seemed to be absent of any emotion. He did not express sadness, impatience, or frustration during any of the time Damasio was with him. Damasio observed other patients with similar brain injuries and found they were all lacking emotion and had similar difficulty in making decisions.

The part of the brain Elliot had had removed was the orbitofrontal cortex (OFC). As it turns out, this small piece of brain tissue serves as a connection between the amygdala and several cortical regions of the brain. The OFC serves to integrate our emotional response with rational cognition. Sometimes, we fly off the handle emotionally, but generally our stream of conscious thought keeps our emotions in control. However, based on what Damasio and others have demonstrated, this stream of conscious thought also needs that "gut feeling" in order to decide. Apparently, we cannot buy a TV based solely on a rational decision, we also must have an emotional nudge; there is always something we like a little better in one choice than the other. Emotions are not necessary for right or wrong answers, i.e., 2 + 2 = ?. Elliot, and others like him, did not show any diminishment in intelligence, they were still just as smart, could still be objective. Yet, they could not be subjective. Subjectivity does not necessarily have a right or wrong answer; it requires an opinion, a feeling. To make a subjective decision requires emotional input. Without emotional input, Elliot has a difficult time deciding if he wants chicken or fish.

Real world decisions cannot be made without emotion. Thus, Spock's world could not have functioned. The Vulcans had long ago done away with emotion because it was irrational and inefficient. But, as we have seen, this would have only allowed them to make objective decisions. Would not the decision to pursue space travel have to have been a subjective one? For that matter, would not the decision to be only logical also be subjective? Given what Damasio has shown us here it would be an emotional decision to decide to act only rational, and if you were only rational you could not make subjective choices. The Vulcan society would have had a very difficult time progressing on logical thought alone. Gene Roddenberry (creator of *Star Trek*) can, of course, be forgiven, the role of the amygdala was not yet known. Of course, you could argue the Vulcan brain is wired differently and evolution has greatly diminished the role of the amygdala in Vulcan brain processing.

My Amygdala Made Me Do It

Now, knowing what we do about the amygdala and its effect on our emotions and sexual motivation, let's go back to Mike and Molly. Initially, Mike was focused on food and filling his plate with the snacks made available (hunger and appetite are also amygdala-influenced activities). Then a very attractive woman is saying "Hi." Mike's amygdala quickly takes in the scene and goes into excitatory mode. Before his cortex has a chance to consider the ramifications, Mike's amygdala is responding. One thing Mike's amygdala knows (likely from years of emotional memory) is introducing another woman as your girlfriend will end most chances of hooking up with the new one (as it turns out, not introducing her as your girlfriend will likely end your chances of hooking up with anyone). Remember, Mike's amygdala is appetitive regarding sexual motivation and, as we have discussed, responds quickly to sexual cues. At that very moment, Mike isn't thinking with anything but his penis. Well, basically he's thinking with his amygdala, but at this moment it's the same thing. Sure, once Mike has time to think about it, his cortex ruminates a little bit and he experiences Molly's response; he realizes how hurtful and insensitive he has been regarding Molly's emotions.

Does this excuse Mike's behavior, just because his brain wiring and chemistry may cause certain actions? Would that not depend on who is judging him? However, with this knowledge, Mike may become more aware of his behavior and Molly may be more understanding. Such that, the next time he is in this situation, having filed away an emotional memory of Molly's response from last time, Mike resists his first inclination and proudly turns and introduces Molly as his girlfriend.

Texas Clock Tower Massacre

It might be difficult to accept that our behavior can be so completely controlled by our cellular activity, but consider for a moment the repercussions when cells go bad. There is lots of evidence demonstrating how small changes in cell activity can lead to dramatic circumstances for the organism. On August 1, 1966, Charles Whitman climbed to the top of the University of Texas Clock Tower, barricaded the doors, and started shooting pedestrians below. On that day, before he was gunned down by Austin police officers, Charles Whitman killed 14 people and wounded 38. Earlier, he had brutally murdered his mother and wife.

By all accounts, Charles had been a model American boy. He was exceptionally intelligent with an IQ of 138 (99.4 percentile). He was one of the youngest Eagle Scouts in the world at age 12, played high school sports, and joined the Marines after graduating high school. For the most part, Charles was well-liked and described as an outstanding person and a great guy. But, shortly after joining the Marines he started to experience bouts of anger. He began to suffer from severe headaches and, as he described to a psychiatrist, periodic and uncontrollable violent impulses. Charles began to over eat and take excessive notes, writing down his thoughts on everything from his love for his wife, his disdain for the Marines, and his need to control his anger. Charles knew something was not right because in his suicide note he wrote:

> I don't understand what it is that compels me to type this letter. Perhaps it is to leave some vague reason for the actions I have recently performed. I don't really understand myself these days. I am supposed to be an average reasonable intelligent young man. However lately (I can't recall when it started) I have been a victim of many unusual and irrational thoughts. These thoughts constantly occur and it requires a tremendous mental effort to concentrate on useful and progressive tasks [...] I consulted a Dr. Cochrum at the University Health Center and asked him to recommend someone I could consult with about some psychiatric disorder I felt I had [...] After my death I wish that an autopsy be performed to see if there is any visible physical disorder . . .[180]

The requested autopsy revealed that Charles Whitman had developed a "glioblastoma mulitiforme tumor, the size of a walnut, erupting from beneath the thalamus, impacting the hypothalamus, extending into the temporal lobe and compressing the amygdaloid nucleus."[181] As we have seen, the amygdala performs a primary role in the processing and memory of emotional reactions. Additionally, the amygdala has been shown to be heavily involved in the maintenance of behavioral responsiveness even in the absence of an immediately tangible or visible objective or stimulus.[182] It is likely this tumor put pressure on Charles' amygdala region, causing sustained stimulation, a probable cause for the headaches Charles suffered.

This tumor caused more than just headaches. In studies where the amygdala receives continuous stimulation, the subject will experience weariness, fear and rage. These reactions can last for several minutes after the stimulation is withdrawn; ". . . rage and attack will persist well beyond the termination of the electrical stimulation of the amygdala."[183] Other abnormalities observed associated with disturbances of the amygdala include Hypergraphia excessive writing, hyperreligiousness—increased food consumption, inability to sleep, and mania.[184] All are characteristics expressed in Charles Whitman's behavior months before the tragic event. Clearly, Charles' behavior was greatly influenced by the growth of cancerous cells in his brain. The pressure these cells placed on the healthy cells of the amygdala lead to behavior, that despite the efforts of Charles' rational brain, he eventually could not control. Although, his actions were planned and deliberate, given what we now know of his condition, it is hard to say he was acting of his own free-will. It would seem his behavior was being determined by his cellular biology. Despite what we might call a rational effort to override these violent impulses, Charles was unable to defeat the overpowering influence of the amygdala stimulation he was receiving as a result of his tumor.

FIGURE 13: Charles Whitman, 1963. A brain tumor pressing on his amygdala turned an Eagle Scout and Marine into the Texas Tower Shooter.

This section has focused on the discussion of the brain and the role memory and emotion have on our behavior. It is important to recognize that the memory and the emotion influencing our behavior are happening at the cellular level. Our memories are a matter of cellular activity resulting from experiences. Our emotions are a matter of cellular activity resulting from incoming perception. It is these two factors, memory and emotion, operating at the cellular level that will determine our behavior. We behave, react to our environment, based on our cellular biology.

PART THREE

CELLULAR DETERMINISM

CHAPTER 12

Behavior

"Your beliefs become your thoughts;
your thoughts become your words;
your words become your actions;
your actions become your habits;
your habits become your values;
your values become your destiny!"

—MAHATMA GANDHI

How We Act

One evening my daughter and I were having a conversation at the dinner table. The subject was about domestic abuse and when, or if, someone can redeem themselves and be forgiven for his actions. As we were debating back and forth the volume of our conversation began to rise. Then my daughter began to tear up as she started to get upset. I didn't understand, "Why are you getting upset?" I asked, "Because you're yelling at her," my wife said. "I'm not yelling! I'm just talking loudly." Well, my daughter got up and walked out of the room, which I mistook as stomping out of the room, which caused me to complain about not being able to have a simple conversation with anyone. To which my wife responded, "Then stop yelling." "There is a difference between speaking loudly and yelling, and right now I'm yelling, but before, I was just speaking loudly!" Then I stormed out of the room.

After I had had a chance to cool down, I realized it was all very silly. So I apologized to my daughter for raising my voice (in my defense, my daughter is quite the loud talker herself). She told me she wasn't upset with the volume as much as she was frustrated by what she perceived as me not hearing her. She had been debating the issue in school all day and was already tired and moody. I too, had been given some bad news that day; a real estate appraisal had come in well below what I knew to be the true value of a property. It was so low, I was indignant. So my mood, too, had been affected by events from earlier in the day. Therefore, when my daughter and I began our conversation we were already under the influence of emotional events from earlier. Thus, this underlying mood affected both the conversation and the reactions from each of us.

The point of this little story is that neither of us could have behaved any differently that night. I can reflect on what happened and say I should not speak so loudly; I should make sure my daughter understands I hear her point; I should be calmer and less reactionary. And, given this experience, maybe next time I will. But, at that moment I could only have behaved the way I did. My experiences and emotions were determining my actions. I know the argument is "one can choose to behave differently." I say, no, you cannot. In the moment, you are relying on your emotions and experiences to determine how to respond, your responses are immediate (given the restrictions of neurons). These responses are being dictated by the cellular activity of your amygdala, hippocampus, hypothalamus, and cortex.

All of the biological data presented to this point, and everywhere I look, I see evidence that our behavior is nothing more than our cells' response to their environment. Our cells are, structurally and functionally, the result of our genome's and epigenome's response to environmental stimuli. This is really no different than any other cell anywhere in the world. Again, every cell is responding directly to its microenvironment. For multicellular organisms, the microenvironment of the cells are regulated such that, the cells respond in a coordinated fashion. The organism as a whole is in a continuous effort to keep all of its cells responding in a fashion best suited for the survival of the organism. This regulation of the cells happens at the cell membrane interface with the environment—where hormones and neurotransmitters bind to surface receptors. Again, these hormones and neurotransmitters are being released from other cells as a result of environmental cues.

The case I have presented here clearly suggests our behavior is determined by how our cells respond to their environment. Even who we are is a matter of cellular behavior; who we are is the result of two cells, egg and sperm, coming together — something completely influenced by the molecules of the microenvironment guiding these cells to each other. Who we have as parents is also the result of molecular activity at the cellular level. Not only must there be the physical attraction, which stimulates responses from the amygdala, but as we see with MHC molecules, there is also an olfaction component in response to a genetic difference that can influence mate choice. Our genetic makeup is not from choice. We are who we are because of our parents' behavior, as a result of their cellular activity.

This genetic makeup establishes the basis on which the environment will influence development. Immediately upon fertilization, the zygote is being influences by the microenvironment provided by the mother. If she is on drugs or has an infection, the developing embryo will be impacted. There will be changes to the epigenome that will influence the activity of cells as they grow and divide. As we saw with successive male children, by simply having been pregnant before, the mother's environment influences early development that can, and does, greatly affect behavior later in life.

Our memories, too, are determined by cellular activity in response to external stimuli. These memories are established based on what we perceive, or have perceived. All memories are of experiences retained at the cellular level and these retained experiences will influence every decision we make currently and in the future. We cannot make a decision without considering our past; unless of course there is something wrong with our cellular biology.

Our decisions, however, rely on more than just our past experiences; our current emotional state also influences how we perceive incoming stimuli and influences our response. No behavior happens without the combined influence of emotion and experience. No behavior happens without the combined influence of the Now and the Then.

As young children we are primarily influenced by emotion, mainly because we have little experience to draw from. As we age, we rely more heavily on past experiences to moderate our behavior; essentially, we have increased the number of synaptic connections that allow for greater information exchange between various areas of the brain. In both cases, as young children and as older adults, it is cellular activity that is determining behavior. The interconnections between the amygdala, the hippocampus, the hypothalamus, and the cortex interplay to determine how we will respond to any given situation; the greater influence changing as we mature from the amygdala to the cortex. Affect the cellular biology of any of these areas and you will affect behavior.

Social influence

In humans, there is an additional level of complexity that adds to our behavioral responses—the social nature of our species. This adds another element by which cellular activity must account. The social nature of our species means our cells must not only regulate the cells of the organism, but must also be able to interact successfully with other massive collections of cells (other individuals).

This is a point of great importance. Our behavior is and will continue to be influenced by our social interactions of our past and present. In fact, we have evolved to desire, actually need, social interaction in order for proper development. We seek out contact with others. As we interact with others, we develop feelings and opinions about them that will subsequently influence how we behave in their presence. Our behavior is modified based on our knowledge and feelings towards the person with whom we are interacting.

Furthermore, our behavior changes in response to his or her behavior towards us; someone expressing trust in us causes a more generous response from us.[185] Over time, our knowledge of someone may influence our feelings towards that person and so our behavior changes as well. These behaviors are determined by how well our cellular activity responds to the cellular activity of the others (think MHC molecules and how they influence female behavior).

If you consider any argument you have ever had, many times afterward you may think to yourself, "I should have said this." But you didn't. Or, you may think, "I should not have said that." But you did. What determines whether or not you say the appropriate thing? You are in an argument, your amygdala is taking in information from the senses—especially visual cues from the other participant; reading his or her face for emotion. The amygdala's connection with the hypothalamus triggers a hormonal release affecting the entire system, including feeding back to the amygdala. Additionally, your hippocampus is receiving info from these parts of the brain as well as scanning the cortex for past stored experiences. Your response is dependent on the neurons firing at that time, all being influenced by their chemical environment. And, because we are reacting in the moment, our response is the first series of neurons to fire.

Our response feels free. Our will is being expressed. Or is it? Our responses are due to the neuronal activity at the time, cellular biology in response to environmental cues.

Given time to think, we might consider other options; our hormonal level may be different, allowing different neurons to fire. Our hippocampus may find the path to a memory that would have helped make our point. New information, maybe from the argument, may cause a change of opinion. When we are young, much of our actions are influenced by the activity of the amygdala and hypothalamus. As we age, our knowledge base grows (increased synaptic connections) and the cellular activity of the cortex has a greater influence over these other parts of the brain; we appear less emotional. Still, even the knowledge that helps control our emotions is a matter of cellular activity resulting from emotional experiences.

Cells Causing Behavior

Cellular Determinism differs from other types of determinism in that it only refers to the behavior of the organism due to cellular level responses to the environment. It incorporates both the biological aspects of the cell and the physical impact of the environment. This is not a new concept, just a restatement of those before me. "From genes to proteins, from cells to orderly development, from electrical activity to neurotransmitter release, from social communication back to any and all of these levels, we are confronted with a system of somatic selection that is continually subjected to natural selection."[186]

I consider Cellular Determinism a synthesis of several deterministic ideas that, in and of themselves, I find too narrow: Biological/Genetic Determinism, Environmental Determinism, and Social Determinism. Biological Determinism is a more apt term for what I have described here; however, that phrase has already been taken to imply determinism based on a strict genetic influence. Obviously, genes are not the only player and so biological determinism falls short. Environmental Determinism places too much emphasis on the physical surroundings of the organism without giving enough weight to its genetic make-up. As I have already stated, the environment can only influence the genes that are present. Social Determinism is actually a type of environmental determinism, but it places greater emphasis on the interactions between individuals as a driving force in our behavior. Our parents, friends, and society determine who we are and how we behave.

Cellular Determinism allows for, and encompasses, each of the above ideas. It is the genes that will code for the proteins necessary for the cell to respond; it is the environment that will influence which genes are turned on or off; and it is with the people whom we interact in which these two factors are expressed.

Mother's Role

Cellular Determinism seems to place a heavy responsibility on the parents for how their children eventually behave—especially the mother. If our early environment is establishing, to some degree, which genes are on and which genes are off, then clearly this early environment is determined by the mother. As we have seen, having given birth to previous sons seems to influence the internal environment of the mother and, ultimately, the development of successive sons—increased chances of homosexuality. This is the more subtle effects of epigenetics on the cells; there are also the more obvious and well documented cases of alcohol and drug use affecting fetal development as well. Clearly, the environment the mother creates for the developing fetus has great influence over what the child will have to work with in life.

So your mother has a tremendous impact on the functional genes you will have with which the environment can then interact. Note the circularity here, the environment influences which proteins will be available, via the epigenome, for with which the environment then interacts. But your mother doesn't just influence your genome and epigenome make-up. She also is a huge contributor to your memory base in which all of your future decisions will be based. As Roger Waters so aptly describes in his rock opera "The Wall" she puts "her fears into you."[187] This is what all mothers do; now and always. They warn us to stay away from strangers, to look out for cars, or to be careful around fire. All practical messages for survival and certainly information a maturing child should know. But mom also gives us our irrational fears as well. "Be good, or God will punish you." And, she'll give us our biases. "Don't go near hippies they will try to sell you drugs."

This isn't to place blame or fault on mother, remember her actions and behaviors are also determined by her past experiences—her parents. It is a never ending spiral. We are greatly influenced by our mother and her influence then impacts, either negatively or positively, how we raise our children. But how she raises us has everything to do with how she was raised.

Our mother contributes to all aspects of Cellular Determinism. She is responsible for 50% of our genes, she greatly influences our epigenome and she establishes (in most cases) our early social exposure. In none of these influences do we have a choice; yet all will play a role in our behavior. And so it is also true for our mother. Her behavior and how she raised us is a product of her upbringing. This is not to say we will turn out exactly like her, quite the contrary. Her biases may cause us to behave in just the opposite fashion as she might have done. Still, our behavior, either way, will be the result of her influence.

Society's Role

Mom is not the only thing contributing to our memories and experiences. Our environment, in which we are embedded, is primarily a matter of social interaction. These interactions will also contribute to our memories and experiences. Sure, there is the physical element that influences our cellular surrounding (pollutants, temperature, etc.), but our behavior and actions, our experiences and our intentions, are centered on others and their response to us.

We are a social creature like no other and much of this is because of language. Language has contributed greatly to the expansion of our brain and with it an ever increasing complexity to our social behavior. We are not solitary creatures. Sociality is so ingrained that even the act of being solitary is a social behavioral response. Because of this, our behavior and actions are almost completely determined by our interaction with others.

In addition to our mother influencing our memory base, there are those in our early experiences that will also impact our mental and memory development. This can be our father, or our grandparents, or friends our parents had around. To a much lesser extent than our mother, every individual we interact with has the potential of causing a long lasting memory which may influence future behavior. I say "potential" because, obviously, not every interaction is remembered. Events not committed to some kind of long-term memory are not likely to influence future behaviors. In our early mental development, many, if not all, of the memories we establish result from emotional events. If you think about some of your earliest memories, other than your mother, they likely involve some kind of emotion; happy, sad, frightened.

When I was five years old my mother's boyfriend at the time got angry with me for making noise. I was outside playing and having not heeded his request to be quiet, which, frankly I don't remember, he threw a hard soled sandal at me. The trajectory was on target and though I didn't see the sandal, I heard a loud knock as the sandal struck me on the left side of the forehead. The next thing I remember I'm being carried in someone's arms and seeing blood pool in my belly button. My mom's boyfriend had split my scalp open!

I was raced to hospital emergency where a nurse immediately took me from my mother's arms and into the back rooms. I remember thinking, "Wow, we didn't have to wait." All my other previous experiences at the doctor's office seem to have us waiting for very long times. I was placed on a table and a towel with a little opening was placed over my face. I remember watching through this opening the surgeons stitching up my head. It was like a little window and I was able to see the doctors as they worked.

Upon my return home, while my mother was packing, this man who had just cracked my head open, very apologetically brought me a glass of Kool-Aid with ice in it. He then told me if I put my finger in the drink and swirl it around it will cool faster. I don't remember what this man looked like, but in addition to giving me a two inch, crescent shaped scar, he also gave me this little piece of information I have never forgotten. Why? Likely, most assuredly, because learning that swirling ice in your glass cools it faster was tied to an extremely emotional event.

This event then would have a tremendous impact on decisions I would make in the future. For example, my mother must have been embarrassed by the event (maybe by the fact that she would be involved with a man that would do such a thing), because she immediately had me lying about what happened; again, a mother instilling her fears. When people would ask me how I got my scar, I was supposed to tell them, "I fell from a swing and hit my head on a rock." The thing was, I was always being asked about my scar and I was never comfortable telling the lie. Sometimes when asked, I would look to my mother not knowing which story to tell, the swing story or the truth. And it seemed to me the adults knew my story was fabricated. When I got a little older, I purposely started wearing my hair long, especially my bangs, because I could then cover my scar and not have to deal with questions.

This one event had significant influence on my behavior form that time on. Some of that impact has certainly been conscious, like the decision with my hair, and some likely subconscious, an aversion to hard soled sandals. The event triggered the establishment of specific nerve cells (as we have seen with the Jennifer Aniston neuron) programmed to fire when that memory is recalled, as well as a network of cells that are excited in response to the firing of that neuron.

Additionally, we must recognize the stimulus that caused the firing of that specific neuron and corresponding network of neurons also triggered several other closely associated memory networks. The cellular activity involved when experiencing that event again will include the amygdala as part of the emotional response, the cerebral cortex as part of the memory and rational response, and the hippocampus, which will coordinate this information into an action or behavior. In some cases my amygdala may have a greater influence, causing my reaction to appear more emotional and sometimes my cerebral cortex will prevail, causing my reaction to appear more rational.

Sure, it can be argued that this event is now just an experience, a long-term memory, which is drawn upon whenever making a free-willed decision. But how can that be? How much control do I have over which neurons fire and which ones do not? If I have control, how then is it manifested? If these memories and experiences are cellular based, then my reaction to any stimulus must involve one cell or cells exerting influence over others. So my response is a result of the cells' activity—depolarization, neurotransmitter release, protein activation—in response to environmental stimuli. How can I determine which cells fire, even though it is I who am firing the cells? Is it not reasonable to suspect my response is based on those cells that have the greatest influence (by greatest influence I mean more easily triggered, releases large amounts of neurotransmitters, or in some way influences which neuron fires while others do not)?

In his book *Connectome,* Sebastian Seung describes a phenomenon he calls the "weighted voting model."[188] The firing of neurons is dependent on the weighted input from all their synaptic connections. Take our Jennifer Aniston (JA) neuron for instance; in order for it to fire, it must receive input from neurons recognizing the many features of Jennifer Aniston, blue eyes, blond hair, great smile. Those features more closely associated with Jennifer will have stronger synaptic connections to the JA neuron and so their input will be greater. Once enough excitatory neurons have fired, they will cause the spike threshold of the JA neuron to be reached and it will fire. It could be possible that the neurons that recognize Jennifer's eyes have a strong connection with the JA neuron and so just seeing her eyes might be enough to cause the JA neuron to fire. Or, absent input about the eyes, there may be enough other weaker connection, hair, smile, derriere, to eventually cause the spike threshold to be reached and the JA neuron to fire. In the "weighted voting model," a neuron will fire once it has received enough information from all its synaptic connections, some of which will provide a greater influence because of stronger connections.

The firing of any given neuron is the cell responding to input from others cells. I do not choose which neuron fires, as much as the choice is made dependent on external stimuli of the cell. Choice seems to be a matter of responding to the cells that have been excited by the environment.

CHAPTER 13

Quantum Influence?

"The universe is not only queerer than we suppose, but queerer than we can suppose."

—JOHN HALDANE, *Possible Worlds*

Probability Waves

Determinism is not a new idea, many before me have accepted this reality; Rene Descartes, while believing our thoughts were indeterminate, believed the physical world was deterministic. Sir Isaac Newton followed with the science to support such claims. Newton's theory of gravity provided all the evidence philosophers like Pierre-Simon Laplace would need, "All the effects of Nature are only the mathematical consequences of a small number of immutable laws."[189] Albert Einstein was also a determinist, once writing, "In the beginning (if there was such a thing), God created Newton's laws of motion together with the necessary masses and forces. This is all; everything beyond this follows from the development of appropriate mathematical methods by means of deduction."[190] By the turn of the twentieth century science had taken a very deterministic tone.

That all changed in the early 1900s with the discovery of several phenomena, which, when taken together, suggested some weird randomness to the universe. First, Max Planck noted light could not be emitted except in packets. Light, and other waves, were given off in quanta (packets) containing certain amounts of energy depending on the frequency of the wavelength, the higher the frequency, the greater the amount of energy per quantum.

Isaac Newton had also alluded to the possibility of a similar phenomenon when he referred to light as tiny pieces of matter he called corpuscles.

This was soon followed by Einstein's discovery of the photoelectric effect. Einstein demonstrated that these quanta of light (photons) could knock electrons off a metal surface. Einstein went as far as to say all light is composed of photons.

Niels Bohr followed with his description of the structure of an atom, providing an explanation of atomic stability. Bohr suggested that electrons could only occupy certain orbits around the nucleus of the atom and that all matter absorbs and emits radiation in lumps. Bohr thus stated that the electron energy within the atom is quantized and confined to discrete orbits. Bohr was also able to explain why different elements gave off light at certain frequencies; this light corresponded to the energy being emitted from electrons as they dropped to lower orbits.

Then, along came Louis de Broglie. He looked at Planck's idea of light existing as individual packets of energy, Einstein's theory of matter and energy being interchangeable ($E=mc^2$), and Bohr's idea of quantized electrons, and wondered if all matter could exist as a wave. Proof of de Broglie's theory would come in 1927 with slight modification to a classic experiment.

In that year, Clinton Davisson and Lester Germer demonstrated the properties of quantum mechanics in a classic, oft repeated, and simple-to-perform experiment. The phenomenon was first demonstrated by Thomas Young in 1804 when he proved that light behaves as a wave. The experiment is known as the two-slit experiment and can be performed with a flashlight, an index card and a screen or wall. Cut two narrow slits into the index card and shine the flashlight through these slits. What appears on the screen behind is a light and dark band pattern representing where the light is striking the screen and where it is being interfered with. We recognize that light travels in waves and so the corresponding pattern that is observed can be explained using the phenomena observed when two waves combine. Just like when two ripples are created in a pool of water, in some places the crest of one wave will combine with that of another to amplify the wave, and in other places the crest and trough may combine to cancel the wave; thus, the alternating light and dark bands on the screen (Figure 14).

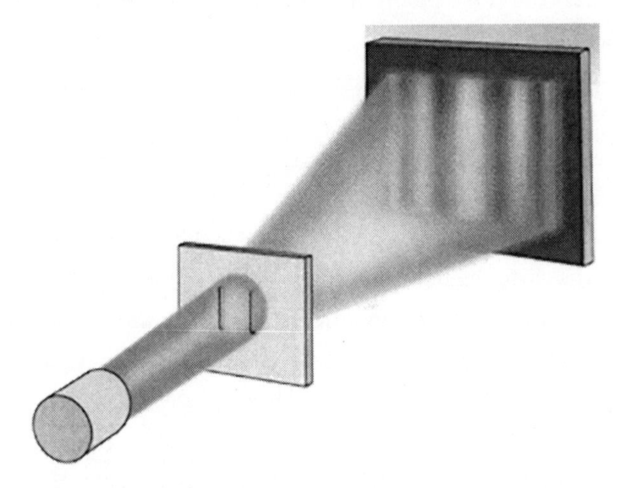

FIGURE 14: Two-slit experiment showing wave properties of light.

However, what Davisson and Germer did, instead of using a beam of light, was to fire a beam of electrons through the dual slits. Surprisingly, this beam of individual electrons was displayed on the screen in the same light and dark bands as seen with the light, indicating that the electrons were behaving as a wave. It appeared that matter, indeed, behaved both as a particle and as a wave. This particle-wave duality for all matter is quantum mechanics. (Figure 15)

FIGURE 15: This screen shows the individual location of electrons from an electron beam after having passed through the dual slits. Clearly, the pattern represents wave properties. The electron fired at the two slits is traveling as a wave until it strikes the screen, at which point its position is measured and its particle location known.

The next observed phenomenon that changed the way we thought about the universe came when Werner Heisenberg described his Uncertainty Principle. Heisenberg realized that if one were to measure the position of a particle, it would require at least one quantum of light. Since these quanta are emitted as waves, one's accuracy will depend on the distance between crests. Heisenberg argued that to increase accuracy in measuring the position of a particle, one then would need to decrease the distance between crests—thus increasing the frequency of the wavelength. But, as one increases this frequency, so too, does the energy of the quanta increase. The increase in energy will have a greater disturbance on the velocity of the particle.

What this implies is, because of the disturbance created on the particle being measured, as we more accurately measure the position of a particle, the less precise we can be about its velocity. To more accurately measure a particle's velocity, we would use a lower frequency wavelength of light, one with less energy, so as to minimize the disturbance; however, this would decrease the certainty of the location of that particle.

Essentially, the Uncertainty Principle states that we can never know the exact position and velocity of any particle; in determining one, you disturb the other. Heisenberg's principle lead him, along with Niels Bohr and others, to formulate the theory that particles do not exist in a defined location at any one time, instead they are in a quanta state of position and velocity.

Given the results of the two-slit experiment, coupled with the Uncertainty Principle, it was clear that electrons, and what has turned out to be particles in general, behave both as a particle and a wave, and we can never know, nor is it knowable, the exact location and velocity of any particle at any given time. This is quantum mechanics and its introduction to the world threw into questioned the validity of determinism. Einstein struggled mightily with the premise of quantum mechanics that there is a randomness and unpredictability to nature, famously stating "God does not play dice." But quantum mechanics has stood the test of time and withstood many scientific challenges. In fact, Einstein himself would bolster the theory with his efforts to dispel it. In the microscopic world, we really do not know where the electrons are, or where the photons are, or where they will be. Instead, we can only speak of potentialities; where they might be or where they may end up.

These realities lead Max Born to propose a probability wave. Particles, instead of being found at any one location, were spread over a wave of probabilities. While there may be greater likelihood of a particle being in one place versus another, there is still the chance, no matter how infinitesimally small, that the particle could be anywhere in the universe. So, while there is a high probability that the electron may strike the screen in a particular location, we cannot say that it definitely will (Figure 16). The particle can be anywhere in the universe until it is observed, at which time its wave function collapses and it coalesces at a single location. Quantum Mechanics is freaky! At the microcosmic level, nothing is anywhere until it is observed.

Field of Action Potentials

Is it possible that free-will is very much like a probability wave? Let us assume you wake one night to the smell of smoke; your senses are detecting, your mind is perceiving, and your memories are being recalled. You have several options at your disposal. You could cower

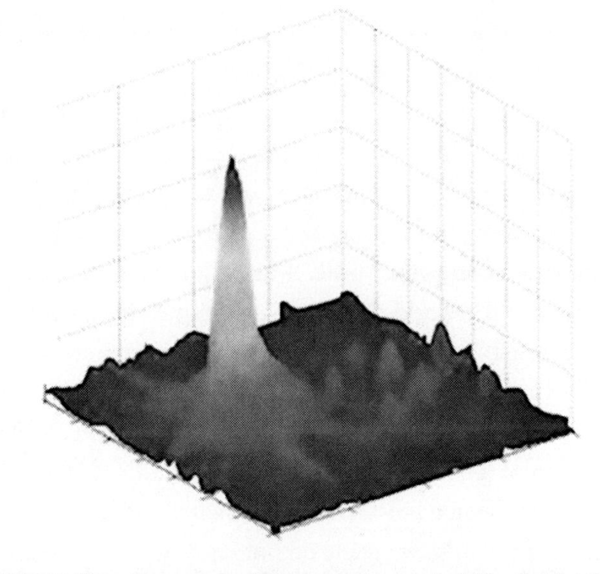

FIGURE 16: This diagram depicts a possible probability distribution for an imagined electron based on its last known location and velocity. The tallest peak represents the most likely position.

under your bed until you burn, or until someone rescues you. You could jump from your bedroom window and go running down the street yelling fire! You could get up and calmly locate the source of the smoke and call 911. These are just a few possible responses. Who knows? Maybe the smell of smoke inspires you to take a shower. Of course, one of these behaviors has a greater likelihood than the others. Are the options available to us similar to the probability wave described in quantum mechanics?

Let's acknowledge, when faced with making a decision, we really have every conceivable action or behavior, based on our experience, available to us; much like an electron having the probability of being anywhere, and everywhere, until observed. With the electron, we can calculate it having a greater probability of being in one place over another with a very strong possibility of being close to its observed location, especially if we have information about its history.

We can do the same with our actions. Given any scenario, like the one above, there are always a tremendous number of possible responses, but there are usually one or two responses that are of much greater likelihood. Are these options then similar to the location of an electron in a probability wave? Instead of a wave however, better we think of our options as a field of possible responses, a probability field of action potential.

Imagine the above scenario and the many potential actions one could take. We can then consider the decision, as demonstrated by the action taken, the result of settling on one of these many possibilities. Applying the principle of quantum mechanics to decision making, we could speculate the likelihood one would call 911 versus the likelihood one would decide to take a shower. Much like the probability wave for a particle—where the particle may have a 90% chance of being in one place, an 8% chance of being another place, a 1% chance of being in yet another location and the rest of the possibility spread over everywhere else in the universe—are our decisions determined similarly? An observer may estimate we have a 90% chance of calling 911, we have an 8% chance of running down the street yelling fire, a 1% chance of cowering under the bed, and a very small, infinitesimally small, chance of taking a shower. And like with the electron, our prediction improves with the more we know about the individual's history.

Knowing the relative position and velocity of an electron can help us predict where it may strike a screen, but we can never be 100% certain. But this uncertainty, at times, can be so small as to not warrant much consideration. Would that not also be true of someone taking a shower while their house burned down? While possible, it is so unlikely as to not be considered probable.

Stephen Hawking in his book *A Briefer History of Time* describes it this way: Imagine using Newtonian physics to calculate a dart hitting the bull's-eye of a target. If you know the speed and angle of trajectory, you can predict with great accuracy where the dart will hit the board. But quantum mechanics says that although you may be right with your predictions every time, you cannot be certain the dart will hit the target every time.

You may throw at the target until the end of the universe and never observe a miss, but the probability would still be there.[191]

With probabilities this small, an observer would be justified in not considering other actions as viable options for one's predicted response. Furthermore, if the observer knows some of the individual's history—had he been in a fire before, was he level-headed or did he have a tendency to panic, does he really like to be clean—the observer can better predict the individual's behavior.

Given any scenario we have a field of action potentials and the decision made is the result of a collapsing wave function at the point of a neuronal excitation. The neuronal network that ultimately leads to the response of the individual is chosen randomly, according to quantum theory. Most of us would probably call 911, a few of us would run down the street yelling fire, and someday, somewhere, someone may wake to his building on fire and decide to take a shower.

There is a certain randomness to decision making. It is not uncommon for people to state they are not sure why they chose one response over another. There are many interviews where the individuals will not be able to articulate why they decided on one action versus another action. However, now consider that we are the observer in our thoughts and behavior. There is a constant stream of self-explanation.[192] Think about the decisions you make. Do you always know why you do what you do, especially, in times of crisis or when pressured to respond quickly? How often have you said to yourself, "Why did I do that?" or "What was I thinking?" Is this the manifestation of quantum mechanics in our decision making?

Does this randomness, provide the unpredictability, that equates to free-will? I am not sure that it does. Randomness does not imply a choice as much as a settling. Free-will implies choice. I freely decided to take a shower while my house burned down. But if this is a response due to the unpredictability of quantum mechanics, it's not then freely chosen; it's settled on randomly.

How is that free-will? If anything, it seems to argue for indeterminism. However, if we accept that this indeterminism could be the root of

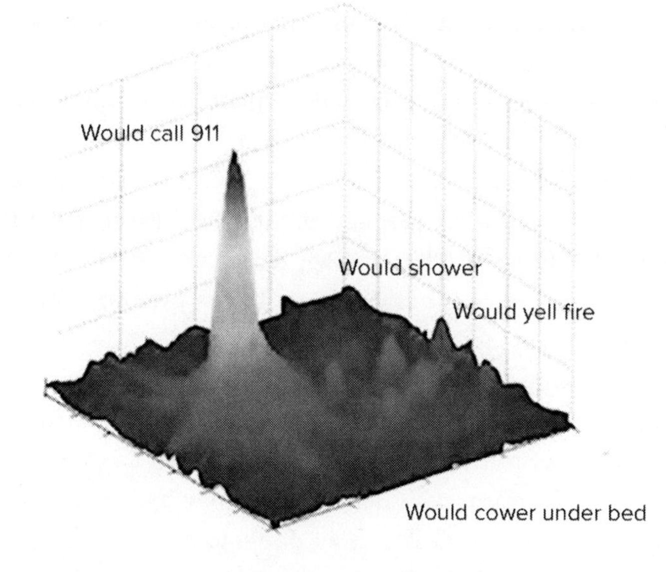

FIGURE 17: Field of action potentials. Imagine this is your probability wave distribution for the fire scenario. All actions are possible, but some will have a greater likelihood.

a free-will decision, would we, society, accept some of the least probable outcomes as free-will? "Mr. Jones, why did you decide to shower while your house was on fire?" "Because, I was sure my neighbors would call 911, and frankly, the fire alarm caused me to crap myself." Would society accept this as a free-will decision, or would we think there is something mentally wrong with this guy. In fact, there are only a few responses we would accept as reasonable, rational responses to this scenario. Anything else would likely be considered a momentary lapse of reasoning, maybe brought on by panic, anxiety, stress, or some other underlying cause. The randomness of quantum mechanics provides for indeterminism in that we cannot know the exact location of particles, but randomness is not choice and therefore cannot be free-will.

Spontaneous Generation

There is one other major departure between quantum mechanics and behavior due to Cellular Determinism. Quantum mechanics allows for particles to spontaneously come into existence, but cells can only come from pre-existing cells. In what is known as quantum fluctuations, particles and antiparticles spontaneously form only to quickly annihilate each other in preserving the law of energy conservation. Laboratory studies have demonstrated quantum fluctuations occur everywhere and all the time. Lawrence Krause, in his book *The Universe from Nothing,* describes this phenomenon this way: empty space is nothing more than a broiling brew of virtual particles, popping in and out of existence.[193]

The same cannot be said for cells. Louis Pasteur demonstrated in his classical flask experiment[194] that cells do not spontaneously generate. Except for the very first cell, all cells come from pre-existing cells. A cell's existence is determined by the biology of the previous generation of cells. Cells cannot pop into existence like particles of quantum fluctuation. So, whereas quantum mechanics allows for randomness, indeterminateness because of quantum fluctuation, no such indeterminateness exists at the cellular level. Cells and their properties are determined by the cells before them. Quantum fluctuations may influence a cell's behavior and that would be determined by how it affected molecules of the cell. But again, such influence, if there were any, would be random. But as I have already stated, randomness does not equate to free-will.

Often in our society, we hear of terrible atrocities that seem improbable, such as random mass shootings at schools, movie theaters, public gatherings. Such acts do not seem to make sense. They appear to be on the same level of improbability as would be taking a shower while your house burns down. Such random or un-determinedness, could result from fluctuations at the quantum level. This implies, while highly improbable, that each of us has the chance to make a decision that would seem to be insane. I reject this premise in that there are certain acts that, except in the case of a biological disorder, I will not do.

Could quantum fluctuations cause such a biological disorder? It is certainly within the realm of possibilities, but such impact would be felt at, and manifested at, the cellular level. Maybe such fluctuations could cause the shutdown of a particular gene through the influence on various molecules like methyl groups and RNA molecules (epigenetic influence). However, behavior, caused by some random event at the quantum level, while suggesting a future that cannot be known, does not necessarily imply free-will.

If randomness is not free-will, and clearly it is not, then what is free-will? Philosophers have been debating this question since man started questioning his actions. More recently, scientists have joined the debate. Most dictionaries and people, in general, define free-will as free choice, a voluntary action made of one's own accord. It is the expression of personal choice above and beyond what might be determined by the physical world, or a divine entity. But how is this choice manifested? Are our actions the expressing of our free-will? Do our thoughts also qualify as free-will or must we act on them? Free-will implies direction. This is why the randomness of quantum fluctuation does not qualify; there is no direction of choice. If my will is free, then I must be able to direct my choice, it cannot be the result of some random phenomenon.

CHAPTER 14

Free-Will

"The I that I am, explains these thoughts to the I that I am."

—RHAWN JOSEPH

My Will is Freely Chosen!

The main issue most people have with cellular determinism is the fact that it removes our own free-will and the personal responsibility of others. From the moment we become aware of our "selfness" in relation to others and the world around us, we feel strongly our decisions are freely made. We control what we do. There is a sense of choice and we, ourselves, decide on that choice. Our actions then, our behaviors, are a matter of us making a cognitively conscious decision on how to respond to any given environmental stimulus. Our self-awareness, the idea of I, essentially dictates a notion of free-will. As soon as we are conscious of our individuality, our thoughts become egocentric: "I" run after the gazelle; "I" find roots and berries; "I" have sex with my mate. The notion of us controlling our own actions seems automatic.

When you consider your thoughts, are they not always in first person? Everything is in relation to us. Are our thoughts not also, generally, in words? We think in language. In fact, language is likely one of the leading contributing factors to the expansion of the cerebral cortex. The language part of our brain provides words to our frontal lobe, which then uses these words in a conscious stream of thought about what is going on, what has been experienced, or what might happen (based on experiences). We are explaining our thoughts to ourselves as we think them.[195]

Often, children will repeat something they have heard or say something inappropriate at the exact wrong time. The child is often seeking information or putting pieces together, "Mommy, is this the lady you said was rude?" It can be quite embarrassing and you wonder to yourself, "Why, would he say that in front of her?" It happens because the child is thinking-out-loud.

At the earliest ages of language development, children three to four years old explain their play to themselves. They are both the speaker and the listener. They explain to themselves what they have done after the fact: "I drew a picture of a rainbow." As we age, our egocentric speech begins earlier in the sequence of events and more internalized.[196] Instead of explaining to ourselves what we have done, we progress to explaining what we are doing and eventually progress to explaining what we are going to do.

Additionally, while the egocentric speech continues, it becomes more and more internalized until we are no longer voluntarily describing our thoughts. We go from telling ourselves out loud, "I drew a picture of a rainbow," to telling ourselves, "I am drawing a picture of a rainbow," to finally, with no verbalization, telling ourselves, "I am going to draw a picture of a rainbow." It is about the onset of puberty that we see neurological activity resembling that of adult activity during language tasks. By this time, egocentric speech, while still self-directed and serving an explanatory function, has become completely internalized.[197]

It is also around puberty that interconnections between the language part of the brain and the frontal lobes, as well as the transmissions from left to right hemisphere, become more complete. Our emotional, visual and motor functions develop well before our language centers have matured, leaving these areas in control of behavior until our frontal lobes can interject their influence.

Clearly, if there is free-will, it must result from our internal thought process being able to exert control over behavior. Does this happen at puberty? Is the internalizing of our thoughts, in conjunction with them explaining what we are going to do rather than what we did, the onset of free-will? If so, then free-will is dependent on the cellular maturation of the brain.

As our mental development progresses, our thoughts become more egocentric. This egocentricity brings with it the sense of "I". With this notion of individuality also comes the sense of control over our actions, and with control, ultimately, the sense of free-will.

I See You Seeing

I have an Australian Shepard that requires me to play with him often. He is high strung and wanting to play ball or Frisbee just about all the time. He is a smart dog, as dogs go, but occasionally he will not see me throw the ball and so does not see where it goes. I'll point, "It's over there." He'll look at me. I say again, "Over there." As I point again. He will inevitably take off in the wrong direction. "No, no, back over this way." I yell. He will come running back to me. So, I point again, "Over there." And he runs off searching again. Eventually he will come across it or I will walk over to it to show him. Pointing is useless and I know it, but I do it anyway.

Pointing is useless because only humans and maybe some primates can gaze monitor, but not dogs. Gaze monitoring is the ability to look at someone's eyes and be able to track their gaze to where it is they are looking. We can look at a pointed finger and follow where it is pointing. Surely you remember the elementary joke where you say, "Hey, look, a dead bird," as you point to the sky; then laugh hysterically as your little sister looks to where your finger is pointing, and you tell her, "Dead birds don't fly, silly."

The ability to gaze monitor develops between eight and twelve months of age.[198] The infant is able to follow one's gaze and then check back and forth to confirm he is looking at the same thing. With this ability we now have an internal sense of what the other person might be experiencing, or at least at what it is they are looking.

Being able to sense what others might be experiencing underscores the second issue people have with determinism, the apparent lack of personal responsibility it seems to imply. People struggle to accept others are not in control of their actions. We want others to be responsible for what they do. We want to hold people accountable. It is important in a socialized species like ours that people do as they say they will. Without responsibility for one's action ethical and moral behavior becomes moot. It would be easy to blame one's behavior on not having control over what happens, being fated to act a certain way. As individuals with a sense of selfness, we also project that sense of selfness onto others, expecting that if we have free-will, so do they.

This ability to project what we consider a sense of self onto others such that they, too, must possess a sense of self, may result from a group of brain cells called "mirror neurons."[199] These cells seem to allow us to interpret what others might be feeling. Mirror neurons are a special group of brain cells that fire when we watch someone perform an action. These neurons, it turns out, also fire when we perform the same action. The idea is these specific cells allow us to simulate the intentions of others and the feelings behind their actions.[200]

One way this is done is by decoding facial expressions. Much in how we communicate our emotions is done through our eyes and face. This system of mirror neurons is activated during facial expression; whether we are making the face, or interpreting the face. If we make a frown or observe a frown, the same mirror neurons will be activated. If we smile at someone, or receive a smile from that person, the same neuron cells will be activated.[201] These cells help us interpret and understand the intention of others' actions. This is why mood can be so contagious.

Being able to interpret others emotions through their facial expressions, along with our unique ability to gaze monitor, puts us in the position to project a sense of selfness onto others. If that person has a sense of "I" like I have a sense of "I," (he must, because we express similar facial expressions to similar emotions) then that person must also be in control of his behavior. He, then, like me, must be responsible for his actions.

It is essential in the social interaction of our species for individuals to be responsible for their actions; it is the only way a leader can demand allegiance or a community impart justice. The social nature of our species has been one of the driving forces in the evolution of our mental capability and consciousness; with this level of consciousness comes a sense of personal responsibility and free-will.

CHAPTER **15**

Free Thoughts, Free Choices?

"To be conscious that we are perceiving... is to be conscious of our own existence"

—ARISTOTLE (384–322 BC)

Consciousness and Free-Will

Can thought alone be free-will? We all have thoughts we do not act on. You could say not acting on a thought is an action. Thoughts are the options, the possibilities available to us. But these thoughts are determined by our brain chemistry and our past and present experiences. Is it the action taken, given these thoughts, that is the manifestation of free-will? Furthermore, whatever action is taken, it must be taken voluntarily. Is it our biologically determined thoughts give rise to a probability field of action potentials, of which by acting on one, we express our free-will?

Thoughts, our thoughts anyway, human thoughts, are generally considered a sign of consciousness. But the idea of what is consciousness is a debate that has been ongoing for centuries. John Locke, in

1690, described consciousness as "the perception that passes in a man's own mind." Rene Descartes proposed a dualistic approach, separating mind (consciousness) from body. Today, neuroscientists tell us consciousness is based on neural events within the brain. In 2004, a group of scientists, led by Richard Frackowiak went as far as to say it is premature to try to define consciousness in any precise way.[202] This said, most of us would agree thoughts are a form of consciousness.

However, is the notion of consciousness as being just thoughts too narrow an interpretation? Most agree consciousness means some level of awareness and usually we are referring to self-awareness. This awareness of self is what we consider consciousness and with this self-awareness comes an innate sense of free-will. But if we consider that consciousness can exist at various levels—mainly as a result of the degree of awareness—then we realize all cells have some degree of consciousness.

Awareness (consciousness) is simply a matter of detecting one's environment. When someone is not able to detect their surroundings, we define him as being unconscious; he is unaware of what is happening around him. Although it doesn't require a response, the only way we know an organism is aware is to measure some kind of response. All responses to stimuli are initiated at the cellular level and by the 'will' of the cells. Response to stimuli equals awareness of environment—equals consciousness.

Furthermore, this 'awareness' is taking place at the interface, indeed, because of the interface, between cell membrane and environment. The simplest bacterium is 'aware' of its environment. They respond accordingly, even predictably, to their environment. As we have already discussed, the cell membrane, with its different surface proteins, binds to and reacts to, molecules immediately surrounding it. This awareness of its microenvironment is no different than the awareness any cell has, anywhere, for their surroundings.

The degree of consciousness a bacterium might have is not at the same level as ours (so we assume), as it is limited by a lack of complexity. Bacteria may be aware of the conditions around them and most assuredly act on them, but we would not consider this awareness any more than responsiveness. The microbes, as best we know, have no recollection of past events and do not plan behavior to achieve a specific goal. However, the cell membrane will have proteins that are a reflection of the environment the microbe has experienced. So, in this way, we could say there is a sort of memory that the microbe carries. The level of consciousness experienced by bacteria would be of the most fundamental order.

As cells began to accumulate into more complex organisms the degree of consciousness also becomes more complex. Multicellular organisms consist of cells, which while still fundamentally responding to their microenvironment, are now communicating to and with other cells. A single cell, living independently, is aware of its immediate surroundings egocentrically. On the other hand, cells of multicellular organisms, while still only detecting molecules in their immediate surroundings, will also be detecting molecules from distant cells. This will allow for a communication with more distant cells, but it also allows any specific cell to be aware of the condition of this particular collection of cells (the organism as a whole). The degree of awareness or consciousness is greater because of the additional communication and coordination between cells.

As we move up the animal kingdom, cells begin to specialize. Communication by way of simple diffusion of molecules, while sufficient for cells near each other, and small collections of cells, now becomes too slow for larger collections of cells to respond adequately to their environment. These larger collections of cells must now act in a coordinated fashion in order to respond sufficiently to their environment. Thus, we see the evolution of specialized cells which allow for a more rapid communication between very distant cells as well as a more coordinated response to things going on around the organism. These specialized cells are nerve cells.

As we have already seen, nerve cells are pretty similar, structurally and functionally, across the animal kingdom. Because of these cells, the organism as a whole is now more aware of its surroundings; the organism is perceiving much more information. With this increased awareness comes a heightened degree of consciousness. As we increase cellular complexity so do we increase the level of consciousness.

Over time these new, specialized, cells increase in number and density. Specific structures form consisting of high concentrations of nerve cells, ganglions, and eventually a brain. These new structures are now allowing for a much greater perception of the environment and a more directed, coordinated response. The organism's awareness is such that it can respond with what we might describe as intent. Some actions no longer seem purely reactive, but designed to achieve an end. The consciousness of these organisms is of a higher degree than those before it, simply because they are aware of much more.

I have two dogs, one is a black lab and the other the aforementioned Australian Shepard. I have watched the lab manipulate the Shepard to achieve an end many times. If the Shepard is on the dog bed and the lab wants to lay on it, the lab will go to the dog door and stick her head out as if something of interest is going on. The Shepard, inevitably, will get up to see what it is and the lab will promptly take his place on the dog bed. I know you might say I'm reading more into this behavior than is actually there, it may be coincidence, but such behavior would not be witnessed in a sea anemone or a flatworm. My point here, if consciousness is a measure of awareness, then clearly, as we move up the animal kingdom, as awareness of the environment increases, internally and externally, so does the level of consciousness. Mammals and birds have a much greater level of consciousness than do jellyfish or earthworms.

This brings us to the level of consciousness attributed to humans, self-awareness. With self-awareness we are aware of our individuality, separate from the environment and others. Self-awareness also implies knowing our feelings and traits and even our behavior. As soon as we wake in the morning we are aware of our selfness, where we are, how we feel. Immediately, what seems to be a continuous stream of thoughts turns to either what we need to do, or what we have recently been doing. This is what we know to be our consciousness. Upon inspection we see it is simply just a matter of an increased sense of awareness: awareness of our environment, of our body, of our actions, and of our thoughts.

Humans are not the only animals that seem to possess some level of self-awareness; several studies have demonstrated awareness of self in dolphins, apes, and even a few birds.[203, 204, 205] Studies of these sorts use what is called the mirror test.[206] A dot, or some other marker, is place on the animal without the animals knowledge (presumably when it is asleep or distracted). Then the animal is shown its reflection in a mirror—realize these animals must have prior experience with mirrors—if it recognizes the spot or marker as something new or different (this is measured by the animal touching the spot) then such recognition is attributed to self-awareness. As mentioned, dolphins, all of the non-human great apes (bonobos, chimps, orangutans, and gorillas), elephants and some species of birds (magpies and pigeons) demonstrate some level of self-awareness.[207] We can only conclude then that their degree of consciousness is on a higher level than most animals. Do great apes contemplate their predicament? Do they desire a specific future?

Since we have defined consciousness as being aware, we must realize then any of our past experiences, those we are able to recall anyway, are now part of our consciousness. (I am not going to make a distinction between conscious and sub-conscious here; they are both various degrees of awareness and so contribute to our overall consciousness.) People we know, loved ones who have passed, an unrequited love, are all now part of our consciousness.

What does this mean—to have people and events as part of our consciousness? Simply, at the time the event was being experienced your brain was busy synthesizing the proteins that would lead to the strengthening of synaptic connections, which, in turn, established a neuronal network allowing for a continuous awareness of these people, places, and things. By continuous I do not mean to suggest these memories occupy every moment of thought, rather, for the most part, the memory can be conjured. Being aware of events and experiences adds them to your consciousness where they will influence any decisions made.

Social Consciousness

There is yet another, higher degree of consciousness we must consider, which can best be described as social consciousness. One of the things you may have noticed about the animals who exhibit levels of self-awareness is that they are primarily very social. If you remember our earlier discussion, the flip-side of self-awareness is awareness of others, social awareness. Distinguishing ourselves from others allows for an identity, but also fosters the sense of identity in others. This creates an awareness of others and the notion that they likely think and feel as I do. Empathy develops. From a collective awareness comes what can best be described as a social consciousness. It is the community's awareness that establishes this greater consciousness.

Social awareness is the result of our ability to put ourselves in the place of others. Recall our discussion on mirror neurons; at a very early age we are able to imagine what others are feeling based on our experiences. We cringe when someone is about to get hit, we withdraw an arm watching someone else's arm get smacked, we smile when others smile. The empathy we experience contributes to our social awareness.

Interestingly, studies by Paul J. Zak, demonstrate that the neurochemical oxytocin plays a significant role in establishing bonds and relationships with others. This molecule appears to facilitate many of the behaviors necessary for social interactions: bonding, empathy, trust; with it, an increased social awareness and an increased social consciousness.

Zak refers to oxytocin as the moral molecule; I think it can as easily be referred to as the social molecule. We see increases in oxytocin levels when people have been shown trust, when people share genuine hugs (even with strangers), when mothers are bonding with their children. Oxytocin makes us more trusting, more caring, and more generous.

Furthermore, Zak states the behavior can be turned on and off simply by applying oxytocin to a subject's nasal passages. Oxytocin—a biologically derived molecule, being coded for by the DNA of the cell, binding to receptors on neighboring and more distant cells—changes behavior. Our behavior can fluctuate solely on the presence or absence of this molecule, which in and of itself is dependent on cellular activity. Is our generosity to others really a matter of free choice or are we simply responding to the cellular release of oxytocin, stimulated by some environmental event?

Recently, some physicists have promoted the idea that there is a universal consciousness, one that is shared between all particles of the universe.[208] Such theory uses quantum mechanics and its non-locality feature as the mechanism for such possibilities. Particles that have been in contact with each other become entangled; entangled particles remain aware of what each other does, regardless of any distance of separation. This awareness is immediate and instant.

Yes, it sounds farfetched, but the data is there. Because all particles were at one time in contact with each other, because of the Big Bang, they are all entangled. Therefore, this awareness shared between particles establishes a universal consciousness.

Although, I can accept the quirkiness of quantum mechanics at the subatomic level, I am hard pressed to see how it can serve as a mechanism to influence behavior. If such mechanism is determined, it will likely be a random influence at the microscopic interface between neurons and have little to do with decision making or free-will.

Given the hierarchy of consciousness, where would we see free-will emerge? When would consciousness begin to include free-will? One could argue all organisms respond based on their will: their will to survive, their will to reproduce. If we mean free-will based on voluntary choice, at what point do we consider choice voluntary? Certainly, many mammals and birds have mating rituals in which the female is making a choice. She is choosing among males with the intent of finding the best fit with which to mate. Is this a degree of free-will? Her choice, while certainly driven by biology, does seem to meet the criterion in that it appears voluntary with an eye towards outcome.

In each of the cases discussed above—from the single cell organism to the self-aware mammals and birds—the awareness, and the continuing complexity of awareness, coincides with cellular awareness of the microenvironment with ever increasing complexity. In all cases, the biology of the cell determines what the cell is aware of and, therefore, what the organism is conscious of. Again, this consciousness is a matter of biological activity; if free-will is hidden in our conscious decisions then it, too, must be determined by biological activity.

If I can influence the firing of neurons such that I have free-will, there must be a mechanism permitting this control. What allows for free-will? What actions are free?

CHAPTER **16**

What Is Not Free-Will?

"You are who you are when nobody's watching"

—STEPHEN FRY

Childhood

In our efforts to identify free-will, it might be more productive to identify what society would agree is not free-will. Randomness cannot be free-will—by definition free-will requires direction. Where else can we agree behavior is not being controlled by free-will? Childhood?

We would all agree, and the science seems to support, that the brains of children are not developed enough to be held accountable for their behaviors. Society has already acknowledged as much and we do not punish children for crimes committed the same way we do adults. In fact, we are much more quickly to come to the aid of a child than an adult. We rationalize that children are less likely responsible for their predicaments, or least did not know better.

On the other hand, adults should know better, we are much more quick to judge an adult's predicament before assisting than we are with children. Adults are responsible, while children are still maturing. Studies of the developing brain also show as much: interconnections between the two halves are not complete, brain wave function is immature, and identity of self is still being established.[209] Although there is no defined age, it appears the first signs of brain maturity that may account for free-will appear around puberty.

Emotion

Emotion also seems to override our ability to decide freely. Again, society has accepted as much. We acknowledge that people can become so angry or distraught with emotion that they behave irrationally. While held accountable, it is often to a lesser extent. We have all done things in a purely emotional outburst where we might later wonder "What was I thinking?" If free-will is a conscious, decision making process, in which pure emotion can impair or mask, then behavior under such conditions cannot be free-will. Animals, we have already established, show emotion on some level, yet we would not consider behavior form these emotional states free-will. Still, it can be difficult to determine at what point one's behavior is solely controlled by emotion and to what degree society feels one should be held accountable.

Consider the murder case of Ron Goldman and Nicole Brown Simpson. By all intents and purposes, it looks like a case of passion. While a jury of his peers found O.J. Simpson innocent of these murders, Vincent Bugliosi, in his book *Outrage: The Five Reasons OJ Simpson Got Away with Murder*, does an excellent job of detailing Simpson's guilt; sometimes the prosecutors' blow-it.[210] Simpson had a temper, which had been well-documented. People with tempers can blow-up in a violent rage that often has them regretting their actions. Most of us have experienced moments where our temper gets the better of us and we may say or do things we would not otherwise do. In moments of rage like these, it can feel like we have no control. Couple this with intense jealousy, another emotion most of us have experienced at some point, and Simpson was a powder keg that night.

This is not to absolve Simpson of his crime by claiming it was out of his control, it is to explain that his cellular biology, based on the conditions of the moment, led to his violent behavior. I do not think he freely murdered Ron Goldman and Nicole Simpson, as much as he reacted to a situation. I think this is borne out in his behavior after the fact: the slow chase down the L.A. freeway, the threat of suicide, the stalemate with police. These may be seen as the behavior predictable of a guilty man who is being controlled by a flood of stress hormones.

This is also not to say O.J. Simpson should not be punished. Part of our experience is the society in which we are embedded and so society (think environment) will react deterministically, as well. Society will punish this behavior, regardless of the underlying cause. This is similar to the behavior of an immune cell; it will destroy a cancerous cell regardless of the cause for the cell becoming cancerous.

Studies have shown that people will experience activation of reward areas in the brain when punishment is delivered to a non-cooperative player.[211] Brain scans have revealed that dopamine-rich areas of the brain become activated when punishment is being delivered. We get pleasure out of punishing the bad guy.

Evolution has built in a cellular response to "fair" behavior as a result of the selection pressures created by our social nature. Humans needed each other to survive, which requires reciprocity; without it, the whole troop could be in danger. Those unwilling to play fair, while gaining a momentary advantage, would likely find themselves with no one to play with. The pleasure we get from punishing bad guys is a necessary social response. In part it is due to not wanting those things done to us and not wanting someone to "get away" with something we could not. Our cellular biology enforces our moral nature.

Let's note here that the horror of this act now becomes an experience for all those familiar with the case. It is part of the collective memory of society, part of our social consciousness, and yes, it will determine cellular activity that will ultimately determine the behavior of those exposed to these stimuli.

Addiction

Addictions are another one of those conditions by which society recognizes one's judgment is impaired. Alcohol, cocaine, LSD all can significantly impact our behavior and influence our decision-making process. The impairment of one's judgment is sometimes used to excuse one's behavior: "I'm sorry, honey; I was so drunk I don't remember." But, while we may excuse behavior while impaired, we absolutely hold one responsible for the actions taken to become impaired. We might be able to understand one's behavior because that person was drunk, but we blame him for getting drunk in the first place. However, with addictions, even before the impairment happens, the addiction has already had an influence on behavior. At what point does the addiction interfere with being able to choose freely?

When we introduce drugs to the microenvironment of our cells, these chemicals are going to interfere with our neurotransmitters in one way or another. They may bind to receptor sites causing excitation, or they may bind to neurotransmitters themselves causing inhibition. There are any number of possible interactions between the cells and these chemicals, all leading to some sort of cellular activity, or inactivity, which produces a behavior. But as I mentioned above, our behavior is actually modified even before the drug is taken. With an addiction, the individual will adjust his actions to get the drug.

In one way or another, we are all addicted to something. We should realize we are not just talking about addictions to illegal drugs, but addictions to caffeine, nicotine, prescription drugs, runner's high and even falling in love. Any chemical, or chemical release, can greatly influence our behavior. But our efforts to achieve that altered state are also influenced by the addiction. If you are a coffee drinker, your addiction to caffeine almost completely controls your morning routine. I have traveled with individuals who, if we don't find them a cup of coffee first thing, their surliness can almost be unbearable. If one doesn't get that cup of coffee then a grumpy individual greets the day.

Nicotine is another strong addictant, as anyone trying to quit smoking can attest. Methamphetamines are so powerful, for some individuals, a single use is enough to cause a person to become addicted. The amygdala region of our brain, involved in emotion and pleasure, has the greatest concentration of opiate receptors; this can facilitate the development of a heroin addiction. Crack can block oxytocin receptors causing a mother to neglect her children. As Robert Palmer stated, some of us are addicted to love. The dopamine, vasopressin, oxytocin release from someone liking you can be intoxicating and addicting. Do you know people always looking for love? Once addicted, our free-will can become impaired.

But, one can argue, these addictions are self-inflicted and are the result of choices freely made before the dependency, before the altered behavior to achieve the high, before judgment was impaired and while free-will was still operating. The first time one smokes that cigarette or joint, the first time one drinks that beer, the first time one snorts that line, those choices were being freely made before the influence of the chemical. Except, it is unlikely the choices are, indeed, being freely made.

In some cases, certain individuals have a genetic predisposition to addiction and addictive behavior. Their addiction may only require the initial exposure, and by this, I mean the environment in which the opportunity to do the drug presents itself. In other cases, peer pressure, pressure to fit in, desire to fit in, all can lead to a behavior that ultimately was not freely chosen, but rather induced by intense social pressure.

The impact being social beings has on determining our behavior is significant. Prior to that initial opportunity to do drugs, our mindset towards drugs has already been established by the environment and ideas we have been exposed to early in life. In addition, the willingness to participate will be greatly influenced by those around us at the time. We cannot remove ourselves from the influence of our social surroundings.

When I was thirteen, I spent a summer hanging out with my cousin Shane. He was a couple years older than me and was from the "bad" side of the family. The boys from Shane's mother seem to always be in trouble. One day Shane and I went to one of his friend's house. There were three guys and a girl, all older than me by a couple years. I watched as they rolled a joint and started passing it around. It was offered to me and I declined. I had always been told how bad drugs were and I remember my mom telling me to stay away from the "pot-smoking hippies" that lived a few apartments down. So, there was no way I was going to smoke and I didn't.

I hung out with these guys for the rest of the day and night and they seemed to have a pretty good time. I didn't see anything terrible happening.

Even though I didn't smoke pot on this occasion, the experience did have an impact. Like I said, these guys seemed to have had a good time without any drama. This was counter to what I had observed with alcohol. My mom and step-dad had always told me how bad pot was, but my step-dad would drink a six-pack of Olympia beer a night. By the fourth one, he would become verbally abusive. If you could keep him under three he was fine, but you could never keep him under three. This was without fail. I had observed this behavior change not just with my step-dad, but with others who drank alcohol.

A couple months later, Shane was staying at his brother's apartment near where I lived. One weekend we got together to hangout. Shane had met a couple of girls earlier, so we walked over to their house. While we were sitting around, Shane pulled out a joint, lit it, and handed it to one of the girls. She took a hit and passed it to her friend, who took a hit then passed it to me. I had not yet smoked, although I had been around it a couple times. Shane looked at me to see if I was going to take it, or wave it off. I might have waved it off except—there were girls here, pretty girls—and they were smoking, and I wanted to be cool, and I wanted the girls to like me. So, I took the joint put it to my lips and inhaled. Was this decision made freely? Looking back, it seems the circumstance—pretty girls—determined my action. Like

Bill Clinton, I didn't inhale much smoke; at least it didn't seem to me that I did. From what I recall I did not get high.

The stage had been set. The next summer, my best friend Roger stole some pot from his mother and a bunch of us boys, maybe six of us, sat around one warm evening smoking it. This was the first time I recall getting high and I liked it. You can argue that I freely choose to start smoking pot but when I reflect on my actions of the time, I'm not sure how I could have decided any other way. The lies I had been told about marijuana and the pot-smoking hippies (who had always been very kind to me), coupled with my observations of people who drank alcohol, greatly influenced my evolving opinion of pot. Then, put in the position where peers are now pressuring me—trust me, for a young insecure boy just the presence of a pretty girl is pressure—I succumbed.

In each situation, my action, my behavior, was predicated on, determined by, the events prior. The first time I'm with Shane I said no because that is where my knowledge and experience is at that time. Drugs are bad. The second time I'm with Shane, the presence of the girls and the knowledge of previous experiences leads to saying yes. Finally, having already smoked with those girls, it was easy to say yes when Roger and my friends began smoking it the next summer.

In each case, my decisions were limited by my knowledge and experience up to that time. Lack of knowledge and experience is another way of saying one does not have the neuronal make-up, the correct synapses, to include other options as potential choices. We will revisit the implications lack of knowledge has on free-will. Here, suffice to say, one's decisions are only going to be based on the neurons and synapses established to that point.

Given the above, I think it can be argued that drug addiction interferes with free-will. Furthermore, even the initial act of doing drugs is more a product of social influences than it is a matter of choice. It might be easy to discount drug addicts as weak, or people that give into social pressure as weak, but the truth is we all give into social pressures. Much of our behavior is a result of social pressure; the need to conform to another's expectations, or society's rules, or a government's laws.

Is succumbing to these pressures a free-will choice or a matter of one's experiences and influences? If your upbringing has been in a very drug friendly, drug tolerant environment, is society surprised when you either do or deal drugs? In fact, you may be told at a young age drug dealing is the family business, it puts food on the table, it is a noble enterprise. And why should you not believe it? These words come from trusted friends and family. Any choice we make, whether it is to experiment with drugs or not, whether it is to smoke or not, to eat fast food or not, depends entirely on our experiences and knowledge.

Ignorance

Does being ignorant also affect our free-will? Does one's lack of knowledge impact one's choices? Absolutely! How can it not? Your choices, decisions, can only be made based on the information you have at the time of the decision. You may often hear the lament, "If I only knew then what I know now," implying that different decisions would have been made given this additional information. Of course decisions would have been different; the knowledge base—the cellular activity—would have been different. What we learn changes the cellular structure of our neurons, creates and strengthens synapses, thus changing the options from which we have to choose.

If I believe homosexuality is a choice, based on my upbringing and the influences in my life, then I might view such behavior as deviant and perverse. Given this view, I might choose to limit certain rights of this group of people by voting for or against various propositions. However, as my knowledge of genetics grows by way of courses and study, I may realize choice may not be the appropriate term. My views may change based on this new knowledge as well as my subsequent decisions.

This is not much different than past practices of hanging witches or using leaches and bloodletting as medical treatment. In each of these cases we may not fault the individuals, saying instead they did not know better; they lacked knowledge to make an informed decision. Society can be tolerant of ignorance.

But is ignorance a restriction on free-will? I think it is. If free-will, as defined, means free choice, then, if your free choice is limited by your lack of knowledge, so is your free-will. Free-will is impaired by ignorance.

Brain Abnormalities

What about brain abnormalities? Do they interfere with free-will? Well, this is a no brainer, of course they do. Most of us would agree that damage to the brain can cause abnormal behaviors for which the individual would not necessarily be held accountable.

Split Brain Syndrome, where the individual cannot control the actions of his right hand, might be an example. The particular patient in question has, embarrassingly, grabbed the breasts of nurses and other women.[212] Normally such actions would land him a lawsuit, but it is understood he cannot control that part of his brain. Turrets Syndrome is another example of normally unacceptable behavior being tolerated because of an understood brain condition. These, are obvious abnormalities. What about developmental differences that are more subtle yet just as influencing on behavior?

Autism, a genetic condition by all accounts, is a disease in which the child's neuronal development impacts his ability to develop socially. Individuals with autism lack the ability to intuit what other people are feeling. One of our most human characteristics is our ability to interact socially and to empathize and sympathize with other individuals. In fact, we are often judged on how well we are socially adjusted; parents strive to have their children interact well with others. Although not understood, autism interferes with the development of these social skills and abilities. We do not fault autistic people for their behavior; we understand their emotional detachment is a result of a developmental abnormality in the brain and so accept peculiarities we might not tolerate from other people.

Interestingly, studies have shown what can best be described as an autism gradient among men and the science disciplines.[213] The brain is hard-wired differently in men than women, especially with regards to the amygdala and its connections to other parts of the brain.[214] Women, in general, are more emotional than men and this difference is borne out in structural differences in the amygdala nuclei. Men have a more muted emotional response, also a result of hard wiring. When one examines the qualities and characteristics of people with autism (and this is just a list of social skills), it is easy to see men share more of these qualities than do women:

- Little or no eye contact
- Resistance to being held or touched
- Lack of knowledge of personal space
- Responds to, but does not initiate social interactions
- Makes honest but inappropriate observations
- Does not share experiences or observations with others
- Difficulty understanding rules of conversation
- Difficulty in maintaining friendships
- Difficulty in group interactions
- Talks excessively on one or two subjects
- Aversions to answering questions about self
- Unaware or disinterested in events going on around them
- Prefers to be left alone[215]

Right out of the gate, men, in general, have more autistic characteristics than do women. I myself, exhibit many of these traits. I rarely initiate social interactions, occasionally make inappropriate observations, and fixate on a specific subject to the annoyance of my wife and daughter. If you are male, I'm sure you may see some of these qualities in yourself, and if you are female I'm sure you know men, maybe all the men you know, who exhibit many of these characteristics. Evidence suggests these qualities are due to structural differences in brain development and this difference in brain development has a strong genetic component.[216]

As we move up the spectrum of more analytical professions from accountant to scientist, we see men with more and more of these autistic characteristics. There is a gradation even among the sciences with biologists expressing some autistic characteristics, chemists more, then physicists, mathematicians and engineers the most.[217] It seems the more analytical one is, the more autistic traits they exhibit. We often judge people on their social skills and ability to interact with others, yet these abilities (our behaviors) are being determined by the hardwiring (nerve cells and connections) of the brain. Is a physicist anymore to blame for stating, "Those jeans seem to enhance the increasing expansive mass of your gluteus maximus," than an autistic person would be?

Another piece of information is worth noting. In one of the studies conducted by Simon Baron-Cohen that demonstrated the relationship between autism and families with fathers in the analytical fields, an interesting piece of data emerged. There appears to be a significantly higher number of manic depressants in the liberal arts while significantly fewer in the sciences.[218] Manic depression is an emotional state. In drawing a very broad conclusion, this fits in line with what I would expect when cellular activity is determining behavior. I would describe the fields of the liberal arts as more emotional. The individuals in these fields are probably more influenced by emotion and so we should not be surprised to see a greater incidence of emotional conditions in the families of students in liberal arts studies.

Finally, we cannot forget the environment likely also plays a role. A new study suggests a mother's antibodies may also be linked to the development of autism.[219] We should not be surprised that in addition to gene input, autism may also be influenced by the cells' surroundings. These studies underscore the importance and significance genes and environment play in determining our ability to interact with others, and, ultimately, our behavior.

So, obviously, we excuse erratic, irrational behavior when there is a clear "brain abnormality." When is behavior a result of a brain abnormality and when is it a matter of free-choice? Are autistic people responding freely? If so, for what do we hold them accountable? If not, then we have to ask, are engineers responding freely? Given the many characteristic similarities between autism and engineering, where do we draw the line of excusing one's behavior? I have heard many the one liner among biologists, "You'll have to excuse him, he is an engineer," in reference to peculiar behavior from my more analytical colleagues. Consider Sheldon of *The Big Bang Theory*,[220] a theoretical physicist with numerous quirks: his social skills leave much to be desired and so, much of his behavior is tolerated, if not excused, by his companions.

This line between accountability and pardon is very much dependent on the individual and the situation. The more knowledge you have of someone, both of their mental ability and of their past experience, the better you understand their behavior.

Sanity

As a society it seems we accept insanity as being biologically determined. Sometimes environment is accepted as playing a roll, but biology dominates. It is accepted that an insane person must have a chemical imbalance or a neuronal abnormality causing his condition. "People in their 'right mind' don't act that way!" We don't hold insane people to the same level of "responsibleness" as we might people deemed sane. We even let people claim temporary insanity, a condition brought on by biology and/or environment, as a means of not being accountable for one's actions.

So we acknowledge insanity is a biologically determined phenomenon, then by extension, must sanity also be a biologically determined phenomenon? Are we saying that insane people do not have free-will, at least at the moment of insanity, but sane people retain free-will? If a biological disorder, or environmental exposure, can cause insanity, therefore taking one's free-will, then free-will must be biological.

If insanity, however defined, can take our free-will, then insane people live a deterministic life. At what point does a sane person lose free-will because of disease, environment, or whatever? Does free-will only equate to being able to make rational decisions? If insane people don't have free-will because they behave irrationally then are we defining irrational behavior as deterministic? And, if irrational behavior is deterministic then why isn't rational behavior deterministic? If insane people are not responsible because of determinism then how can sane people be responsible if their lives are also determined?

Accepting that cellular activity, or problems with normal functioning of cellular activity, can cause a person to be insane, commit unspeakable acts of violence, or behave in ways that are anti-social, we must also accept that cellular activity can cause a person's sanity, 'speakable' acts of kindness and gregarious social behavior.

Preferences

Let us turn our attention to preferences—our likes and dislikes. Surely, there must be a manifestation of free-will when we consider what pleases us and what does not. Nothing is more representative of who we are than what we like, right?

Preferences do define us in many ways, however, they seem to be more genetically and culturally (environmentally) based than being a matter of choice. Some of the things we like are a result of our genes and a preference for those sensations; for instance, the preference of more dissimilar MHC molecules by women. On the other hand, many of our likes are established by our early experiences. What we like to eat is generally determined by what we are fed; there is a reason most of us like our mother's cooking.

As a young boy in the '70s I watched a lot of afterschool TV and situational comedies. When I could, I preferred to watch the shows with pretty girls. While I didn't think it mattered to me much at the time, I did have a tendency to prefer one type of woman more than another. One of the shows I watched was *Gilligan's Island*—it aired in rerun every afternoon. If you are familiar with the show you know two of the characters were Ginger, a tall, beautiful redhead, and Mary Ann, a shorter, pretty, brunette. Ginger was the sexy one and Mary Ann the wholesome one. For some reason, I always found Mary Ann to be the more alluring.

Three's Company, was another sitcom I was watching at this impressionable age. The story is of a man living with two lovely ladies. Again, the women were contrasted with Chrissy, a tall, beautiful, blonde, and Janet, a short, pretty brunette. As you might have guessed, my preference was for Janet. Physically, I found her more appealing, preferring the shape of her hips and rear. Additionally, both Mary Ann from *Gilligan's Island* and Janet were portrayed as the smarter of the two, the more feisty, and the more independent. Ginger and Chrissy were the sexier, more simple-minded of the pairs. Which of the qualities attracted me more—the physical appearance of Mary Ann and Janet, or their personalities? It was likely a combination of the two.

I tell you this because of the similarities these two women share with my wife. My wife does not only resemble these women in physical appearance; she is short, pretty, and brunette; she also has a personality similar in many respects to these characters: smart, feisty, and independent. My preference did not preclude me from dating women with different physical and personality traits, but in hindsight, many of the women I dated did fit this description. Even the little girl I stole a kiss from in third grade was short, pretty, and brunette. I remember thinking the first time I saw her how pretty she was; there was no one in class I thought prettier. Given the very early age this preference was established, it seems unrealistic to consider it one of choice; rather, it seems clear it was determined by genetics.

As I mentioned earlier, I don't think we have any real control over who we find attractive. You like who you like and hope they like you back. We also have no control over the environmental preferences established early in life, the things to which we are exposed. We like certain types of foods because that is what we were fed. We have certain beliefs, biases and prejudices because of the situations and events we have experienced. We don't know why we like certain things, we just do. Why do some prefer chocolate ice cream to vanilla ice cream? It's a matter of preference, but it is not a matter of choice and not a matter of free-will.

Given what we have discussed here, it would seem much of our behavior is a matter of cellular responses to the environment more than it is what one could consider choice. Children lack the brain development to have free-will and, by extension anyone with similar degrees of brain development. People overcome with emotion, people on drugs, responses to peer pressure, lack of knowledge, brain abnormalities, and preferences all seem to be behaviors less of choice and more of circumstance. Accepting then that any, or all, of these behaviors are more a matter of cellular responses determined by environmental conditions than actions being voluntarily taken, greatly restricts the parameters for which free-will can exist. Still, if none of these apply and a person is acting with all their capacities intact, then their behavior might qualify as being freely chosen.

Surely, the feeling of being in control of one's actions is innate; our responses are voluntary based on our preferences. But how is this voluntary response generated, what is the mechanism that allows for choice? Where is free-will?

CHAPTER **17**

Where is Free-Will?

"Consciousness is a transparent brain representation of the world from a privileged egocentric perspective."

—ARNOLD TREHUB

Brainiac

I have seen it described that the conscious mind is but a very small percentage of the overall brain activity and that most of our neuronal goings on are of the subconscious (unconscious) mind.[221, 222] Free-will, because it is voluntary, has to be of the conscious mind. Therefore, the above would suggest free-will is the result of brain activity in a small part of the brain. If we could define specific areas of brain activity to be conscious thought, say activity in certain areas of the cerebral cortex, and then were able to pinpoint behavior to brain activity in these areas, we could theoretically conclude then that whatever act undertaken in which these areas of the brain are stimulated, must be of free-will.

Sebastian Sueng described "Connectomes" as being maps of neuronal networks; connections between neurons identified such that each stimulus activates an assembly of cells.[223] If each thought, or recalled experience, activates a specific cell assembly, how do we describe the action that follows? The excitation of a specific cell assembly, which leads then to a specific action, implies cellular determinism.

One could argue, instead of a deterministic mechanism, the activity in the cerebral cortex, being rational rather than emotional, voluntarily led to the corresponding action; activity in the cerebral cortex suggest conscious thought went into the response. Is it enough to identify activity in the cerebral cortex as conscious, voluntary ponderings, which lead to freely chosen actions? No cell assembly would be the same from one moment to the next, each would change subtly, based on the excitation from the previous experience. Even though there may be a Jennifer Aniston neuron with an assembly of neurons that fire in leading to the excitation of that specific neuron and an assembly of cells that fire after that specific neuron is stimulated, with each new event the cell assemblies will vary slightly.

Could it be that the subtle changes taking place in these cell assemblies represents the manifestation of free-will? A recent study demonstrated that a brain cell assemblage in the frontal cortex fires about 1500 ms before the decision to perform an action is made.[224] We must first think of something before we can react. What triggers the firing of those cell assemblages? Cellular environment.

Imagine trying a criminal for murder and being able to image the brain as he is being questioned by the prosecutor. If brain activity occurs in those areas where we say conscious thought takes place, could we then find the suspect guilty of his own free-will?

Then, however, would the contrary also be true? Imagine being able to see OJ Simpson's mind at the time of his questioning. What if instead of having brain activity in the conscious areas of the brain, we saw an overwhelming amount of activity in the emotional centers and subconscious regions? Even though the genetic evidence strongly suggests he is guilty of the crime, would these brain scans have revealed a moment of emotionally charged, jealous rage mainly outside our defined area of conscious mind. Would we have to find him not guilty by reason of temporary insanity because he wasn't using the part of the brain that allows for free-will?

No Will or Two Wills?

In a paper published in the *Journal of Cosmology,* Dr. Rhawn Joseph identifies specific parts of the brain that he says seem to contain our free-will.[225] Using individuals who have suffered damage to the brain due to trauma or stroke, Joseph has determined that free-will resides in the medial regions on the frontal lobes. Joseph defines free-will as ". . . the ability to make plans, consider alternatives and choose among them and act on them."[226] When these abilities are lost, so then is free-will. As examples of lost free-will, Joseph describes locked-in syndrome, a condition in which a person may be completely paralyzed yet still be fully conscious and aware. In this case, the inability to act leads to a loss of free-will.

However, when we examine Dr. Joseph's subject closer—journalist Jean-Dominique Bauby, who had suffered a stroke in 1995 leaving him completely paralyzed—we see that he may indeed have continued to exercise free-will, even though his body could not respond. Though Bauby could not move his body, he was able to communicate by blinking. Over a two year period he learned to blink the alphabet and dictated his memoirs, *The Diving Bell and the Butterfly,*[227] to his caretakers. It seems reasonable to argue that Bauby was still exercising free-will by choosing to learn the alphabet and taking the time to tell his story. It is true he could not act on all of his thoughts due to physical limits, but isn't that the case for everyone? Certainly, there are things we would like to do that we physically cannot; many of us may want to be able to hit a fastball, carry a note, even fly. Not being able to do so does not take away our free-will. No, physical constraints do not take away our free-will as much as they define the parameters by which it can operate. Somebody who has lost their legs would not be described as having lost their free-will to run, only their ability to run.

Joseph also describes a condition in which individuals act against their free-will. In circumstances where the supplemental motor areas (SMA) have been severed from the medial frontal lobes of the brain, the patient's primary and secondary motor areas may act counter to the patient's will. In these cases, an arm, hand, or leg may act completely independently of the conscious mind. The patients may grab any nearby object, a pen, glasses, a hammer, and start using it against the desires of the patient themselves.[228] In these patients where the connection between the decision making part of the brain and the acting part of the brain (SMA) has been compromised, the behavior of parts of the body seem to be determined and controlled by external stimuli.[229] The environment of these patients determines their actions. They can't stop from picking up the pen, the glasses, or the hammer.

In cases where the corpus callosum has been severed, patients behave as if they have two separate free-wills. In the extreme case known as Alien Hand syndrome, the left hand will act and behave as if it has a mind of its own. In one case, a patient's left hand would pull a lit cigarette from his mouth even though the patient wanted to smoke. Another patient would reach out and grab the breast and buttocks, much to his own embarrassment, of nurses, doctors, people within reach.[230, 231] As best as could be determined, the left side, no longer having a connection with the right side, began thinking on its own. As Joseph describes, "Surgical destruction of the neural pathways linking the right and left supplemental motor areas and medial frontal lobes will result in two independent streams of mental activity."[232] Joseph assigns these two streams of mental activity as each being a free-will.

Circumstances such as this, where it appears a patient has two consciousness', or two wills as Joseph describes, fits well with our discussion of consciousness. Consciousness is a matter of awareness and so both halves of the brain, while no longer communicating, are still aware of the sensory input coming in; and so, respond to this input. However, the inability of the cells to communicate, caused by the severing of the nerve fibers between the SMA and the medial frontal lobes, diminishes the awareness of the individual. With a diminished awareness comes a diminished consciousness. The actions of these individuals suggest they have two wills, but what we are really seeing is two systems of awareness responding to the environmental stimuli before them. The lack of communication between cells prevents the sharing of information and thus consciousness, too, is diminished.

Accepting Joseph's definition of free-will, "the ability to make plans, consider alternatives and choose among them and act on them," the evidence he presents regarding its location in the medial frontal lobe, with the necessary supplemental motor areas, supports the idea of free-will being cellular based. As Joseph demonstrates, damage to certain cells interferes with one's ability to carry out those features defined as being free-will, or creates a second free-will with both seemingly to meet the criteria of free-will. However, this is only an illusion of free-will. In both cases, the responses being observed are a result of the patient's brain cells reacting to the environment being presented. "Locked-in" syndrome, Alien Hand syndrome, etc., are behaviors and actions (or lack of actions) determined by cellular activity; in these cases, damage to cellular activity.

Brain Waves

Can we measure free-will by measuring brain waves? If we accept that free-will is the result of a certain level of cognitive ability, then monitoring brains waves could serve as a way of determining competence. We know that infants and young children do not generate the kind of brain waves associated with higher thinking.[233, 234] From birth to the age of two, the brain is pretty much operating in a delta wave state. Other brain waves are present, but the predominant are delta waves. These waves are associated with the deepest sleep in adults. All mammals exhibit delta wave activity and maybe all animals do.

Clearly, it is difficult to say whether a baby has free-will, or just lacks the ability to express its free-will. If we conclude that delta wave brain activity is not associated with cognitive thought, in the sense that decisions can be made with a sense of repercussions, then we have to feel free-will comes from a more complete development of the brain and brain wave activity.

From the ages of two-to-four, we start to see increased theta wave activity in the brain. This kind of activity seems to be associated with motor activity development, seeking, finding, jumping, running. After the age of six, we start to see increased alpha wave activity and at about the age of twelve we begin to experience sustained periods of beta wave activity. Alpha waves are associated with a state of calm consciousness, while beta waves are indicative of an active or focused consciousness. Is it here, with the higher brainwave frequencies that we start to see the emergence of behavior we can describe as free-will?

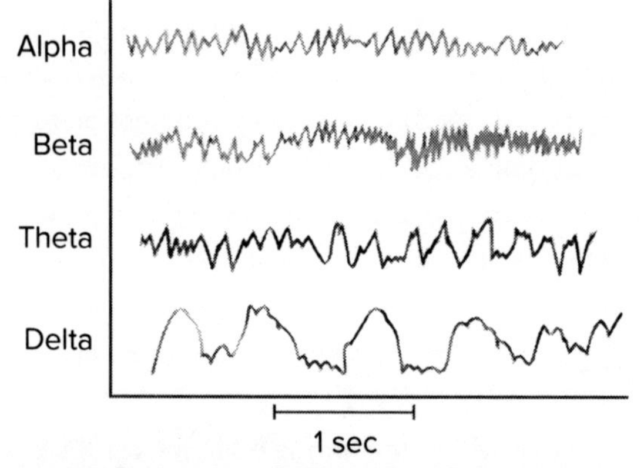

FIGURE 18: Brain waves: 0–2yrs mostly Delta waves, 2–4yrs sees increased theta activity, at 6yrs increased Alpha waves and around 12yrs substantial Beta wave activity.

Society accepts that at a certain point children are not developed enough to be completely held responsible for every act they do. We consider these first few years a learning period. But at some point we expect people, even our children, to become responsible for their actions; responsible because they have free-will.

But if free-will is a level of development, we should be able to define when it appears. Alpha and beta wave activity corresponds with certain levels of consciousness, levels which are reached only after sufficient brain development. This suggests there must be a critical mass of synaptic connections, coupled with the right mixture of neurotransmitters, before we have an appropriate level of consciousness and self-awareness—and thus free-will.

Split Brain Studies

Rhawn Joseph has also conducted several studies on split-brain functioning in children; examining the level of communication between the two hemispheres of the brain at various ages of development.[235] It has been shown that the corpus callosum is not fully myelinated until near the end of the first decade of life.[236] Myelination of the axon nerve fibers passing between the two hemispheres allows for faster and more complete communication between the halves of the brain. The left half is the language side of the brain and so things coming into the right side have to be relayed to the left side before one can express what they have experienced.

Separating children into three groups of ages four, seven, and ten, Joseph conducted a series of experiments to see how well information was relayed between halves of the brain. The children were shown a picture (for instance, a girl blowing out a candle) to either the left or right hemisphere, then the children were asked to describe what they saw. Children in the four-year age group had difficulty transferring all the information; they might have recognized there was a girl in the picture, but when asked what she was doing they would not be sure and so filled the gaps in with made-up information.

For instance, instead of seeing a girl blowing out a candle, the children in youngest age group often reported seeing a girl, but when asked what she was doing they would mistakenly report, opening a gift, or eating cake, or some other activity. Only part of the information these children had observed had been transferred to the left side; they filled in the gaps with information based on what they had experienced from similar events. The right side of the brain sees a girl blowing out a candle, but all the right side can relay to the left side is a girl, maybe in a birthday setting. When asked what she is doing, the left side responds with an answer that makes sense based on the information received from the right side. If it is a girl at a birthday setting, the left brain answers the question by filling in the gaps of information with material from past experiences. If she is at a birthday then she is eating cake, or opening a present.[237]

Why would the children lie about what they had seen? Why not just say, "I don't know."? Well, they are not really lying and they don't know they don't know. The left brain took the information coming in from the right, as incomplete as it might have been, and integrated it into what the child already knew, from experiences, to form a complete picture of what was going on. So, when the child answered, "eating cake" when the girl was really blowing out a candle, the child believes she is telling the truth.

In the seven-year old children, the incidence of fibbing declined and by ten years old, the incidence of fibbing had essentially disappeared. As the children became more mentally developed, as the connections between the halves of the brain became more myelinated, the transfer of information became more complete. There was now no need to fill in gaps with made up information because the children could now access all the information available.

These studies clearly show how cellular development and activity are necessary for the development of free-will. In Joseph's definition of free-will, one has to be able to make plans and consider alternatives. While children can do some of these things, on a limited basis, no one would argue that at four years old a child has the capacity to choose between alternatives with consideration of possible repercussions, especially after seeing these studies. Any decision the child makes would be based on incomplete information.

Certainly, it is easy to argue that the manifestation of free-will does not occur until full development of the brain and the connection between the halves. Free-will then, as it is being defined by Joseph, must make its appearance somewhere between the ages of eight and twelve, allowing for developmental variances.

Adults, too, do as these children did in relying on our experiences to fill in gaps in information. Adults' information is more complete and the flow of electrical impulses more dynamic, but they still rely on past events when interpreting new experiences. Sometimes, in trying to make sense of the limited information (gaps) we generate amazing ideas.

Even our greatest insight is the result of our experiences. Einstein's Theory of Relativity, while brilliant, could only have come about through knowledge he had gained prior to his insight. The necessary neuronal network, derived from past experiences, had to be in place before such thoughts could materialize. Years of studying physics and observing of natural phenomenon built and established a complex series of connections between neurons such that, when the time was right—i.e. enough knowledge had been gained -, the right combination of neurons fired and a new idea formed. Most definitely it was a completely unique combination—a new assemblage of cells and dendritic connections that lead to this new idea. This new idea, while built on past experiences, fills the gap in information needed to complete the thought.

Most often, as we saw with the children, the notions we use to fill the gaps in information may be inaccurate. Such was the case with Lamarck's idea for the mechanism by which traits can be acquired. Occasionally, these gaps are filled by ideas that turn out to be extremely insightful, for example, Darwin's Theory of Natural Selection.

CHAPTER **18**

Evolution of Free-Will

"People like us, who believe in physics, know that the distinction between past, present, and future is only a stubbornly persistent illusion."

—ALBERT EINSTEIN

Illusionary Beliefs

When considering the notion of Cellular Determinism, I spent a lot of time thinking about the evolution of free-will. What would be the selection pressure for free-will? Is there a free-will gene? We have several examples of illusionary ideas that likely have an evolutionary base; religion being the most prominent of those. We have no physical facts or empirical evidence to support the existence of God—yet, most of the human population has a belief in a deity; not just today, but historically. While some may argue this is cultural, I submit that religion evolved as a necessary evil in the increasing social nature of humans. We needed some way to stay honest when no one else was watching. It is convenient that deities help us explain what we do not understand, but more importantly, they influence our behavior.

A study designed to observe the behavior of children who were told they were being watched by a deity gave some interesting insight. The study, "Princess Alice is watching you,"[238] demonstrates how being watched inhibits cheating. Children were split into two groups ages 5—6, and 8—9, and then told about an invisible "Princess Alice." The children were given a task that was impossible to complete without cheating and then further divided into groups that were watched by an adult, unsupervised, or watched by Princess Alice. The unsupervised group of children was more likely to cheat in order to complete the task than either the group being watched by a real person or the group being watch by an invisible person. In fact, the Princess Alice group performed on par with the group being watched by a real person; belief in a watchful invisible person deterred cheating.

Several other studies have shown similar results with regard to behavior modification associated with the presence of a supernatural agent.[239, 240] Belief in a supernatural agent has the tendency of enhancing socially favorable behavior. Being watched inhibits selfish behavior and to some degree promotes altruism. Such behaviors are essential in the functioning of social groups and would allow then for a splitting of labor and development of trust. Even children skeptical of Princess Alice first had to "confirm their null hypothesis about (her) existence"[241] before they would cheat. Being watched facilitated social behavior.

Free-will is a Spandrel

Free-will, like religion, may very well be the result of pressures established from the continued socialization of the species. It is certainly important for the individual, in a social population, to be responsible. Generally, there is a division of labor in a community: some have to hunt, some have to gather, and others tend to the young. Someone not pulling their weight would impact other members of the community and so there may be a selective pressure on these individuals in that they might be out of favor and be less likely to mate.

However, that seems more of a selection pressure for socializing, but not necessarily one for free-will. Free-will is more of an individual concept applied society-wide. I suggest free-will evolved not as a result of socializing, but as a by-product of the more defining human characteristic of self-awareness.

In 1979 Stephen Jay Gould and Richard Lewontin first proposed the notion that some characteristics of an organism may be the by-product of selection pressures for a completely different trait.[242] Some characteristicss may not have been selected for, as much as selected with. The example that Gould and Lewontin described involved the spandrels of the San Marco Cathedral. These spaces (spandrels) were the result of an architectural requirement in constructing the arches that would support the dome (Figure 19). Later, the spandrels would be used for artistic purposes.

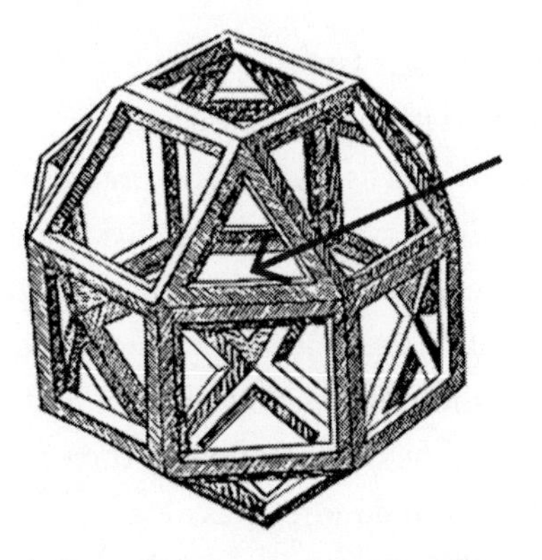

FIGURE 19: Spandrels (arrow)–architectural remnants. The triangular space created by the stacking of squares is a result of the structure rather than an element of the structure.

No architect begins to draw up plans starting with the spandrels; the spandrels were left over spaces that were put to use after the fact. Gould and Lewontin argued that in evolutionary biology, some characteristics and features, might not have resulted from the direct action of natural selection pressures, but rather hitched a ride with some other trait.

Free-will, or the illusion of free-will, is a spandrel. Free-will was not so much selected for as much as it is an indirect result of self-awareness. Given that at some point the neuronal concentration reaches critical mass and an organism becomes self-aware, the next step is free-will. Self-awareness entails a sense of "I." I now feel that my actions and thoughts can be controlled by me. We believe we can do this thing or not do that thing, as though we have a choice, and we can make a decision. It seems that as soon as an organism becomes self-aware then the illusion of free-will must coincide.

It may be possible self-awareness is a spandrel as well. The selection pressures for larger and larger concentrations of neurons (from ganglion to cerebral cortex) led to the evolution of self-awareness, but it would be hard to argue these pressures were selecting for self-awareness, free-will, God, or any of the other notions associated with higher thinking.

Natural selection was favoring greater and greater concentrations of nerve cells because it made the coordination of thousands, then millions, then trillions of cells easier. The organism's behaviors became more elaborate as the nervous system evolved and developed. It's been suggested that the expansion of the cerebral cortex came in stride with our increased social nature. It was an increased collection of nerve cells that aided in the coordination of more and more body cells, the increase in self-awareness that followed then may have aided in the coordination of more and more individuals.

How would self-awareness aid in the coordination of more and more individuals? Along with the idea of self, one realizes then that others are also aware of self. Then we find ourselves wanting others to be responsible. Accountability comes into play. The expansion of the cerebral cortex, because of natural selection pressures due to the ever-increasing societal nature of the species led to self-awareness, free-will, and personal responsibility. All of these may be nothing more than spandrels as a result of the massive increase in brain neurons.

As I finish this passage on spandrels, my mind goes back to a memory of one of my favorite college professors, Dr. Gary Brusca.

As far as I'm concerned, Dr. Brusca, along with his brother Richard, wrote the bible on invertebrate zoology. Dr. Brusca introduced me to the notion of spandrels in a seminar class when he had us read Gould's article. Which, I'll admit at the time, I didn't quite understand. I held to the belief that if a characteristic existed there must have been a selection pressure for that feature; we biologists just had not figured out what that pressure might be. So, in my mind, I imagine Dr. Brusca smiling as I've come to see the point he was making those many years ago.

Of course, this image is only in my mind, a result of neurons stimulated by the word spandrel. Those neurons, reaching out to stimulate other neurons that trigger associated memories of Dr. Brusca and the classroom at Humboldt State University Marine Laboratory. If I let my mind wonder, all sorts of memories return from my days at HSU. All of these memories will be the result of cellular activity: cellular activity now—as I am recalling the memory and cellular activity then—as the memory was being established.

CHAPTER 19

What's Our Reality?

"We must believe in free will,
we have no choice."

—ISAAC BASHEVIS SINGER

The Real Choice

Cellular Determinism versus free-will: which process guides our behavior and which process is real? The answer actually comes down to the definition of reality. Einstein told us everything was relative; time and space are dependent on the positioning of the observer. Personal reality, then, is subjective, each of us with our own reality predicated upon our perceptions.

This is very much in line with Cellular Determinism. Each cell has its own reality based on its interactions with its microenvironment. Each cell has its own consciousness; each cell has its own will—the manifestation of which is in the cell's response to the environment. These cellular interactions with the microenvironment give rise to an organism of which its reality is also subjective, dependent on how well these cells, and by extension the organism, perceive their surroundings.

Quantum mechanics also tells us reality is subjective. Electrons (and other sub-atomic particles) are potentially everywhere until they are measured. They "jump" around with no indication of path. They pop in and out of existence. Such randomness certainly implies a level of indeterminateness. If one does not know where the electron is, how can we predict where it is going to be? Even if we know where it is, in quantum mechanics, we cannot predict with certainty where it will be.

Relativity, quantum mechanics, many of today's interpretations regarding natural phenomenon imply a lack of objective reality. The reality of Newton is no more. Our universe is dependent on a conscious observer, whose presence causes the collapse of these sub-atomic vector states (wave functions), revealing the position of matter. How can we have a determinate system from such apparent randomness? The key to the answer of that question is alluded to in this Stephen Hawking's quote:

> The microscopic world of quantum mechanics is one of determined probabilities. That is, quantum effects rarely alter the predictions of classical mechanics, which are quite accurate (albeit still not perfectly certain) at larger scales. Something as large as an animal cell, then, would be "adequately determined" (even in light of quantum indeterminacy).

At the macroscopic level, the uncertainty of particle location is negligible. It is true, as Hawking described earlier, that a dart we are throwing at the bulls-eye has the potential to end up anywhere. All the particles could jump at the same time and the dart end up on Mars. However, given the number of particles that would have to do that, it is a great statistical improbability. Regardless of what happens at the microscopic world, our ever-day reality is still governed by Newtonian Physics.

So how does this last statement help me reconcile what I know to be behavior by cellular determinism and the illusion I have of free-will? Recognizing reality is subjective, this subjectivity does not only apply to individuals and what they perceive, it also applies to the degree at which perception is sought. Let me explain what I mean by this. From my studies of chemistry and physics, I know the door to my office is composed of atoms and that these atoms are made of a nucleus with protons and neutrons being orbited by electrons. Physics also tells me that the space between nucleus and electron is enormous. One analogy I have seen describes the nucleus as a tennis ball in the center of a tennis court with the electron the size of a grain of sand orbiting at the edge of the tennis court. This means, at least at the atomic level, the door is mostly space and very little matter. This is true for all matter at the quantum level. But when I look at the door, my perception is of a solid object. When I touch the door my perception is of a solid object. When I slam the door the sound gives me the perception of a solid object. At the macroscopic level the door is as solid as if there were no space between subatomic particles. Despite what I may know about the atomic makeup I cannot actually walk through my office door. That is reality.

Is it an objective reality? Well it certainly is at the level at which humans can perceive the environment around them. I can pretty much guarantee you, despite what atomic physics says, that nobody, human, dog, or other disturbance will pass through the doorway without first opening the door. The reality at the classical level can be objective if we provide a narrow description of the system. In this case, a solid door can serve as a barrier to other large organisms.

Let's apply this same analogy now to Cellular Determinism and free-will. Based on my studies in biology, and hopefully much of that background has been presented here, I feel very confident that our behavior is primarily controlled by the activity of our cells. If we look closely enough at behavior, individual behavior, group behavior, societal behavior, we can likely find an underlying cellular cause for the behavior. I get mad, I fall in love, I feel happy, or I feel sad are all conditions brought on by cells releasing hormones and neurotransmitters, as a result of environmental cues, which then bind to receptors on other cells causing the behavior we express. When cells fail to function correctly because of genetics or environment, we see the resulting effects on behavior. It seems behavior, if enough is known, can be explained as a result of Cellular Determinism.

But, when I consider my actions as an individual, as a whole rather than a collection of trillions of cells, then my innate feeling is one of choice. I behave as though I'm making decisions based on rational, cognitive, free choice. These ideals and actions are coming from my brain and my body. Furthermore, I treat people as if their actions and behaviors are also freely made. I expect people to be responsible for what they do, even though Cellular Determinism tells me there is a biological cause for any behavior. It has to be this way; our society could not survive if we did not have a notion of free-will. Free-will serves to enhance our social nature.

Like the solid door, which is not really solid on one level, our sense of free-will is not really free at the cellular level. However, the door is as solid as any I have run into in my life at the macroscopic level and, likewise, free-will is as much of my behavior as any perception I might have. I know it sounds like I'm hedging here, and I probably am, but, even though free-will is an illusion, my reality is I have free-will because that's my perception; just as the door being solid is an illusion, my reality is that it makes a fine barrier.

Our perception is our reality, it does not matter if there is an objective reality. Such a reality will have no bearing on our behavior. We will behave based on the reality we perceive, and it will be subjective and different from every other person. But regardless, the behavior will result from cellular responses to environmental stimuli.

This hedge I take here, with free-will being both an illusion on one hand and a reality on another, is actually not much different than the position taken by many religions around the world, including Christianity.

Christianity tells us God is omniscient, he knows what we will do before we do it, but he also grants us free-will to make our choices. On one level, our actions are determined; they have to be if God knows what we are going to do. On the other level, they are free because we make choices not knowing the outcome. Again, reality becomes subjective. If you are God you know events are determined; if you are human you feel your will is free.

Our cellular biology evolved over millions of years to selective pressures favoring certain responses given specific external stimuli, it should not be surprising when two people respond similarly to these cues. Mike's response when confronted with a pretty girl in front of Molly and my response when introducing my girlfriend in front of potential mates was similar because of our cellular biology and the similarities in the situations. If we could examine what was happening at the cellular level we would likely see certain regions of the brain more active than others, specific levels of neurotransmitters and hormones being released. Given enough information, we could trace the serious of events leading to the resulting behavior.

This is reality at the cellular level. However, both Molly and my future wife expect a certain response based on our ability to make decisions (never mind these decisions are cellular based) deemed more appropriate. If we fail to behave accordingly, then their behavior will adjust as well (it will adjust either way). They expect us to make the right choice based on our free-will. This is reality at the organismal level.

Choices

The first part of this book discussed how we become who we are due to cellular processes. We have no choice in our genetic makeup. We do not choose our parents; their choice to mate and produce us is greatly influenced by their cellular biology. We have seen that our cells can only respond to their microenvironment and that this environment greatly influences the development and behavior of the cell. We have noted the environment can have epigenetic effects significantly affecting behavior now and in future generations. These factors all contribute to our physical make-up, without any decision or choice on our part. Our physical being is solely a matter of cellular determinism.

The second part of the book looked at how our environment greatly influences our brain development, and ultimately how we behave. We discussed decision making (which is all behavior is) as the result of emotional responses to environmental stimuli based on previous experiences. We identified the amygdala as a significant player in our ability to act on incoming information as well as its role in helping us settle on cognitive decisions after weighing pertinent information. We demonstrated decisions are a matter of cellular activity, and when this activity is out of whack, so can be the decision making. Here, again, our choices (behaviors), seems to be nothing more than our cells responding to environmental stimuli.

Finally, in the last section, we examined the concept and implications of Cellular Determinism. We looked for the presence of free-will in our behavior and our anatomy. We introduced the craziness of quantum mechanics and contemplated its effects on decision making and reality. We discussed the potential levels of consciousness and recognized the subjectivity of reality. In the end we acknowledged that free-will, in one sense (at the cellular level), is illusionary, but, in another sense (the organismal level), is as real as our subjective perception allows.

We are merely raw material being shaped by the artist of time; a canvas being painted on, a piece of marble being delicately chiseled. We can no more influence our environment than the canvas can instruct the artist, or the marble direct the sculptor. Our responses to events dictated by the nature of our raw material and the nurture of our environment. We will draw on all our experiences (consciously or subconsciously) to determine our next action.

So, what is one to do with this information? The evidence suggests you do not have free-will, that it is illusionary, and that your behavior and all of our behavior is determined by how our cells respond to their surroundings. If you have no choice, why do anything, why make plans, why set goals, why not just let things happen? Well, they will, regardless of what you do, but you are still a player in the world events. Your actions or inactions will not only affect you, but also everyone who interacts with you and likely people that interact with them. Your behavior will affect others directly and indirectly. To say you have no choice may not be entirely accurate: you still have to make decisions.

I find this philosophy of Cellular Determinism extremely liberating. Recognizing that reality is subjective to each individual, depending on their position and perception, I can be much more accepting when someone's story does not match the events I recall. How many times have you been adamant to your spouse that you gave her some piece of information; i.e., poker game Saturday night, round of golf on Sunday, and she insists you never told her such a thing? You may both be right. Maybe you did tell her, but for whatever reason she did not hear it or it didn't register; her reality is you didn't tell. I know you can argue that there is an objective reality you told her, she didn't hear. But that's irrelevant. Her reality is what she will respond to; her behavior will have nothing to do with your reality. Obviously, you two will share similar experiences and so in some cases there may exist an objective reality for both of you, but it will only be in regards to a very narrow set of conditions. Knowing this, allows me to be much more tolerant of positions counter to mine.

A second empowering feature of Cellular Determinism is the recognition that your presence in someone's environment has the potential of having great impact. People behave based on current environmental conditions in relation to past experiences. Things you say and do can become permanent memories for others. These memories can be drawn on when encountering similar situations. We have the power to influence people's lives in both a positive and a negative way; however, we don't know which way that might be. Still, just knowing your actions and behaviors contribute to the actions and behaviors of others is empowering.

I recognize that what I have told you here is nothing new; many before me have illuminated such ideas. I also recognize that there are other explanations for the behavior we express: free-will, God's will, the Cosmos' will. Fortunately, my subjective reality allows me to accept Cellular Determinism as my objective reality, leaving me to behave according to the environment I encounter.

My behavior will always be such that I am acting based on my free-will and my expectation of others will be that they are responsible for their behavior. It is how it has to be as a result of me being self-aware and recognizing the same in others. However, I know any behavior of mine and of those I observe in others is a direct result of our subjective reality as determined by our cellular biology of the moment. Our environment and genes influence our cells such that their response will determine our free-will behavior.

ENDNOTES

Introduction

1. Mike and Molly, Season 1, Episode 9
2. See any introductory biology text.
3. Again, see any introductory biology or genetics text.

Chapter 1

4. This account for the growth pattern of bacteria in a laboratory setting can be found in many standard microbiology texts.

Chapter 2

5. Watson, Denise M. (2010). *Cancer cells killed Henrietta Lacks, then made her immortal.* The Virginian-Pilot, May 10.
6. Nobel Foundation. (2008). Human Papilloma Virus And Cancer, HIV Discoveries Recognized In 2008 Nobel Prize In Physiology Or Medicine. *ScienceDaily*. www.sciencedaily.com/releases/2008/10/081006093031.htm.
7. From the same article listed in the first endnote for this chapter, in a side panel, an interview with Dianne Daniel, assistant professor in the department of microbiology and molecular cell biology at Eastern Virginia Medical School.
8. Hayflick L & Moorhead P.S. (1961). "The serial cultivation of human diploid cell strains". *Exp Cell Res* 25 (3): 585–621.
9. Olovnikov A.M. (1996). "Telomeres, telomerase and aging: Origin of the theory". *Exp. Gerontol.* 31 (4): 443–448.
10. Myerson, M. (2000). Role of telomerase in normal and cancer cells. *J. Clin Oncol.* Jul; 18 (13):2626–34

11. Takahashi, K., Tanabe, K., Ohnuki, M., Ichisaka, T. Tomoda, K., & Yamanaka, S. (2007). Induction of Pluripotent Stem Cells from Adult Human Fibroblast by Defined Factors. *Cell,* doi:10,1016/j.cell.2007.11.019

12. Vierbuchen, T., Ostermeier, A. Pang, Z.P., Kokubu, Y., Sudhof, T.C. & Wernig, M. (2010). Direct conversion of fibroblast to functional neurons by defined factors. *Nature* 463, 1035–1041 doi:10,1038/nature08797

13. Saenz, A. (2010) *Mouse Skin Into Neurons Without Need for Pluripotent Stem Cells,* Singularity Hub, February 2, 2010 summarizes Vierbuchen et al. work.

14. *ScienceDaily,* (2012). Umbilical Cord Stem Cells Converted Into Brain Support Cells. January 17.

15. Marro, S., Pang, Z.P., Yang, N., Tsai, M., Qu, K., Chang, H.Y., Sudhof, T.C., & Wernig, M. (2011). Direct Lineage Conversion of Terminally Differentiated Hepatocytes to Functional Neurons, *Cell Stem Cell,* doi:10.1016/j.stem.2011.09.002

16. Embryonic germ layers: in the early development of the embryo cells differentiate into three primary germ layers. The cells of these germ layers can now only give rise to specific types of tissues: ectoderm—skin and nerves; endoderm—the digestive tract; mesoderm—muscles and bones.

Chapter 3

17. Of course, this scenario does not describe any real two individuals, but is more a composite of stories heard told over the years.

18. Description of MHC molecules can be found in most college microbiology text.

19. This specific description of MHC molecules comes from the text used in my microbiology course. Nester et al., Microbiology: A Human Perspective, 6th Ed. (2009) The McGraw-Hill Companies

20. Helper T-cells play a major role in regulating our immune response and they do so by recognizing a very specific class of MHC molecules found on cells that present exogenous antigens: B-cells, macrophages and dendritic cells.

21. Wedekind C., Seebeck, T., Bettens, F., & Paepke, A.J. (1995) MHC-Dependent Mate preferences in Humans. *Proc. R. Soc. B.* 260: 245–249. doi10:1098/rsbp1995.0087

22. Roberts S.C., Gosling, L.M., Carter, V. & Petrie, M. (2008) MHC-Correlated odour preference in humans and the use of oral contraceptives. *Proceedings of Biological Science,* Dec 7, 275 (1652): 2715–2722. Published online August 12, doi:10.1098/rsbp.2008.0825.

23. Santos P.S., Schinemann, J.A, Gabardo, J, & Bicalho, Mda G. (2005) New Evidence that the MHC influences odor perception in humans: a study with 58 Southern Brazilian students. *Horm Behav.* Apr;47(4):384–8.

24. It is estimated that only 3% of species are monogamous (and there is doubt about those 3 %.)

25. Garver-Apgar C.E., Gangestad, S. W., Thornhill, R., Miller, R.D. & Olp, J.J. (2006) Major histocompatibility complex alleles, sexual responsivity, and unfaithfulness in romantic couples. *Psychol Sci.* Oct;17(10):830–5

26. *Ibid.*

27. *Ibid.*

28. Roberts S.C., Little, A.C., Gosling, L.M., Jones, B.C., Perrent, D.I., Carter, V. & Petrie, M. (2005) MHC-assortative facial preferences in humans. *Biology Letters* 2005 doi:10.1098/rsbl.2005.0343

29. Annegers L. F. (1989) Patterns of Oral Contraceptives use in the United States. *Rheumatology* doi 10.1093/rheumatology/XXVIII.suppl.1.48–50.

30. Divorceinfo.com (www.divorceinfo.com/statistics.htm)

31. A similar reference to the connection between the pill and divorce rates is mentioned in the book, *Sex at Dawn: How we Mate, Why we stray, and what it means to Modern Relationships.* Ryan, C. & Jethá, C. (2010) Harper Perennial. New York, New York

Chapter 4

32. Taken from "The Bob Newhart Show"

33. Except for the DNA in their mitochondria.

34. Hopper, A.F. & Hart, N.H. (1985) Oxford University Press. Foundations of Animal Development. Second Edition. Oxford, New York. Any general text on embryology or animal development will provide the basic conditions spelled out here.

35. Hormone, cytokine, and neurotransmitter are just different terms for chemical messengers. Each imparts some information to the cell by binding to surface receptors. The various names have to do with their action, structure, target cell, etc., but in the end they are simply chemical messengers.

36. Much of what I describe regarding sperm formation and sertoli cells can be taken from any general text on embryology and development. I got my information from: Foundations of Animal Development. Second Edition. Arthur F. Hopper and Nathan H. Hart (1985) Oxford University Press. Oxford, New York.

37. Studies have shown there seems to be residual mRNA; the mitochondrial machinery of the cell can synthesize proteins to replace degraded proteins using this 'left-over' mRNA. But these residual mRNA molecules are a result of nuclear activity prior to the beginning of meiosis.

38. Griswold, M. (1995) Interactions between Germ Cells and Sertoli Cells in the Testis. *Biology of Reproduction* 52, 211–216

39. Griswold, M. (1998) The central role of Sertoli cells in spermatogenesis. *Seminars in Cell & Developmental Biology*, Vol. 91998:pp 411–416

40. Galdieri, M., Monaco, L. & Stefanini, M. (1994) Secretion of androgen binding protein by Sertoli cells is influenced by contact with germ cells. *Journal of Andrology* Vol.5 Issue 6 409–415

41. Naz, R.K., & Sellamutha, R. (2006). Receptors in Spermatozoa: Are They Real? *Journal of Andrology*, Vol 27, No. 5, September/October doi:10.2164/Jandrol.106000620

42. Brackett, N., Cohen, D.R., Ibrahim,E., Aballa, T.C. & Lynne, C.M. (2007) Neutralization of Cytokine Activity at the Receptor Level Improves Sperm Motility in Men With Spinal Cord Injuries. *Journal of Andrology* Vol. 28, No.5, September/October Copyright American Society of Andrology doi: 10.2164/jandrol.106.002022

43. Eblen, A., Bao, S., Lei, Z.M., Nakajima, S.T. & Rao, C.V. (2001) The Presence of Functional Luteinizing Hormone/Chrorinic Gonadotropin Receptors in Human Sperm. *The Journal of Clinical Endocrinology & Metabolism* vol. 86 no. 6 2643–2648. doi: 10.1210/jc.86.6.2643

44. Shah, C., Modi, D., Gadkar, S., Sachdeva,G. & Puri, C. (2003) Progesterone receptors on human spermatozoa. *Ind J Exp Biol.*. 41:773–780

45. Shah, C., Modi, D., Sachdeva,G., Gadkar, S. & Puri, C. (2005) Coexistence of Intracellular and Membrane Bound Progesterone Receptors in Human Testis. The *Journal of Clinical Endocrinology & Metabolism* vol. 90 no. 1 474–483

46. Lishko P.V., Botchkina I.L., & Kirichok Y. (2011) Female Sex Hormone Progesterone Activates Principal Ca2+Channel of Human Sperm. *Nature* Mar 17; 471(7338): 387–91

47. Bray C., Son, J.H., & Meizel, S. (2002) A Nicotinic Acetylcholine Receptor Is Involved in the Acrosome Reaction of Human Sperm Initiated by Recombinant Human ZP3. *Biol. Reprod.*; 67: 782–788.

48. Meizel S. (2004) The sperm, a neuron with a tail: 'neuronal' receptors in mammalian sperm. *Cambridge Journals: Biological Reviews* , 79:pp713–732 doi:10.1017/S1464793103006407

49. Martín del Rio R. (1981) Gamma-aminobutyric acid system in rat oviduct. *J Biol Chem.* Oct 10;256(19):9816–9.

50. Ritta, M.N., Calamera, J.C, Bas, D.E. (1998) Occurrence of GABA and GABA receptors in human spermatozoa. *Mol Hum Reprod.* Aug;4(8):769–73.

51. Bray, C. Son, J.H., Kumar, P., Harris, J.D. & Meizel, S. (2002) A Role for the Human Sperm Glycine Receptor/Cl2 Channel in the Acrosome Reaction Initiated by Recombinant ZP31 *Biology of Reproduction* 66, 91–97

Chapter 5

52. Stern, S. & Wassarrman, P. M. (1973). Protein synthesis during meiotic maturation of the mammalian oocyte. *J. Cell Biol.* 59, 335a.

53. *Ibid.*, p38, 53.

54. Ginther, O.J., D.R. Bergfelt, L.J. Kulick & K. Kot (2000). Selection of the Dominant Follicle in Cattle: Role of Estradiol. *Biology of Reproduction*, Vol. 63, No. 2, pp. 383–389

55. *Ibid.* p78

56. Cellular organelles responsible for energy production.

57. www.webbooks.com/eLibrary/Medicine/Physiology/Reproductive/Female.htm

58. Cooper, G.M.(2000) The Cell: A Molecular Approach. 2nd edition. Sunderland (MA): Sinauer Associates; Meiosis and Fertilization. Available from: http://www.ncbi.nlm.nih.gov/books/NBK9901/

59. Epel, D. (1979) Experimental analysis of the role of intracellular calcium in the activation of the sea urchin egg at fertilization. In: *The Cell Surface: Mediator of Developmental Processes,* (pp.169–185) S. Subtelny and N. Wessells, (eds). New York: Academic Press. In this paper Dr. Epel demonstrates changes in calcium and pH levels are necessary for continued development after fertilization. We should not expect that these changes would be instant and evenly distributed throughout the egg; the simple process of diffusion will dictate an uneven exposure to any chemical change.

60. Armstrong, D. & Webb, R. (1997). Ovarian follicular dominance: the role of intraovarian growth factors and novel proteins. *Rev Reprod,* Vol. 2, No. 3, pp. 139–146

61. Ginther, O.J., Kot, K., Kulick, L.J. & Wiltbank, M.C. (1997). Emergence and deviation of follicles during the development of follicular waves in cattle. *Theriogenology,* Vol.48, No. 1, pp. 75–87

62. Townson, D. H., & Combelles, C.M.H. Ovarian Follicular Atresia cdn.intechopen.com/pdfs/30307/InTech-Ovarian_follicular_atresia.pdf

63. Wallace, W. & Kelsey, T.W. Human Ovarian Reserve from Conception to the Menopause. *PLoS One.* 2010; 5(1): e8772. Published online 2010 January 27. doi: 10.1371/journal.pone.0008772

Chapter 6

64. Independent assortment describes the process by which when chromosomes align along the equatorial plane of the cell, just before division, they do so independently of what other chromosomes do. This implies then that all genetic variety is possible.

65. Gilula, N.(1979) Cell-to-Cell Communication and Development. In: *The Cell Surface: Mediator of Developmental Processes,* (pp.23–41) S. Subtelny and N. Wessells, (eds.) New York: Academic Press.

66. Sheridan, J.D., Hammer-Wilson, M., Preus, D. & Johnson, R.G. (1978) Quantitative analysis of low-resistance junctions between cultured cells and correlation with gap-junctional areas. *J Cell Biol*, 76:532–544.

67. The inner cell mass consists of cells that will become the person while trophoblast cells are those that become the placenta. Trophoblast cells, while prominent in the early stages of development, make up no part of the person.

68. *Ibid.* 65, p36.

69. Shapiro, B.M.,Schackman, R.W., Gabel, C.A., Foerder, C.A., Farrance, M.L., Eddy, E.M., & Klebanoff, S.J. (1979) Molecular Alterations in Gamete Surfaces During Fertilization and Early Development. In: *The Cell Surface: Mediator of Developmental Processes,* (pp.127–149) S. Subtelny and N. Wessells, (eds.) New York: Academic Press.

70. *Ibid.* 65, p36.

71. Cheng, M. (2007, July 3) Study: Twins form after embryo collapses.

72. Jain, A.K., Prabhakar, S., & Pankanti, S. (2002) *On the similarity of identical twin fingerprints. The Journal of Pattern Recognition,* volume 35 11, November, Pages 2653–2663.

73. Kruszelnicki, K.S. (2004, Nov. 11) Fingerprints Of Twins. News in Science (ABC Science).

74. Spemann, H. and Mangold, H. (1924) Uber Indunktion von Embryonalanlagen durch Implantation artfemder organisatoren. Wilhelm Roux' Arch. Entwicklungsmech. Org. 100:599–638.

75. Hopper, A.F. & Hart, N.H. (1985) Oxford University Press. Foundations of Animal Development. Second Edition. Oxford, New York. 218–222.

Chapter 7

76. Samson M., Libert, F. Doranz, B.J., Rucker, J., Liesnard, C., Farber, C.M., Saragosti, S., Lapoumeroulie, C., Cognaux, J., Forceille, C., Muyldermans, G., Verhofstede, C., Burtonboy, G., Georges, M., Imai, T., Rana, S., Yi, Y., Smyth, R.J., Collman, R.G., Doms, R.W., Vassart, G., Parmentier, M. (1996) Resistance to HIV–1 infection in caucasian individuals bearing mutant alleles of the CCR–5 chemokine receptor gene. *Nature.* 1996 Aug 22;382(6593):722–5.

77. Zimmer, C. (2012) We Are Viral From the Beginning. *Discovery Magazine Blogs*, Friday, 15 at 7:55 AM

78. Horie M., Honda, T., Suzuki, Y, Kobayashi, Y., Daito, T., Oshida, T., Ikuta, K., Jern, P., Gojobori, T., Coffin, J.M. & Tomonaga, K. (2010) Endogenous non-retroviral RNA virus elements in mammalian genomes. *Nature* 463, 84–87 doi:10.1038/nature08695; Received 2 September 2009; Accepted 17 November 2009.

79. *Ibid.*

80. Zimmer, C. (2010, January 12) Hunting Fossil Viruses in Human DNA. *The New York Times.*

81. Matsumoto Y., Hayashi, Y., Omori, H., Honda, T., Daito, T., Horie, M., Ikuta, K., Fujino, K., Nakamura, S., Schneider, U., Chase, G., Yoshimori, T., Schwemmle, M., & Tomonaga, K. (2012) Bornavirus closely associates and segregates with host chromosomes to ensure persistent intranuclear infection. *Cell Host Microbe.* May 17;11(5):492–503. doi: 10.1016/j.chom.2012.04.009

82. Dupressoir A, & Heidmann T. (2011) Syncytins—retroviral envelope genes captured for the benefit of placental development. *Med Sci (Paris).* Feb;27(2):163–9. doi: 10.1051/medsci/2011272163. Epub 2011 Mar 8.

83. Dupressoir, A., Vernochet, C., Bawa, O., Francis Harper, F., Pierron, G., Opolon, P., & Heidmann, T. (2009) Syncytin-A knockout mice demonstrate the critical role in placentation of a fusogenic, endogenous retrovirus-derived, envelope gene. Proc Natl Acad Sci U S A. Jul 21;106(29):12127–32. doi: 10.1073/pnas.0902925106. Epub Jun 29.

84. Heidmann O, Vernochet C, Dupressoir A, & Heidmann T. (2009) Identification of an endogenous retroviral envelope gene with fusogenic activity and placenta-specific expression in the rabbit: a new "syncytin" in a third order of mammals. *Retrovirology.* Nov 27;6:107. doi: 10.1186/1742–4690–6–107.

85. Lamarke, J. B. (1809) *Philosophie zoologique.*

86. Zhang, Y. & Reinberg, D. (2001) Transcription regulation by histone methylation: interplay between different covalent modifications of the core histone tails. *Genes & Dev.*. 15: 2343–2360.

87. Reik W, Dean W, & Walter J. (2001) Epigenetic reprogramming in mammalian development. *Science.* Aug 10;293(5532):1089–93.

88. Iqbal, K., Jin, S.G., Pfeifer, G.P., & Szabo, P.E. (2011). "Reprogramming of the paternal genome upon fertilization involves genome-wide oxidation of 5-methylcytosine". *Proceedings of the National Academy of Sciences* 108 (9): 3642–3647. doi:10.1073/pnas.1014033108. PMC 3048122. PMID 21321204.

89. Phillips, T. (2008) The role of methylation in gene expression. *Nature Educations* 1(1)

90. Cooney, C.A., Dave, A.A. & Wolff, G.L. (2002) Maternal Methyl in Mice Affect Epigenetic Variation and DNA Methylation of Offspring. The *Journal of Nutrition*, vol. 132 no. 8 23935–24005.

91. Weaver, I. Cervoni, C., Champagne, F., D'Alessio, A., Sharma, S., Seckl, J., Dymov, S., Szyf, M. & Meaney, M. (2004) Epigenetic programming by maternal behavior. *Nature Neuroscience* 7, 847—854 Published online: 27 June 2004; doi:10.1038/nn1276.

92. *Ghost in Your Genes*. Transscript; narrator speaking.

93. Krakowski, P., Walker, C., Bremer, A., Baker, A., Ozonoff, S., Hansen, R., & Hertz-Picciotto, I. (2012) Maternal Metabolic Conditions and the Risk for Autism and Other Neurodevelopmental Disorders. Pediatrics. April 9, DOI: 10.1542/peds.2011–2583.

94. Guénard, F., Deshaies, Y., Cianfione, K., Kral, J., Marceau, P. & Vohl, M. (2013) Differential methylation in glucoregulatory genes of offspring born before vs. after maternal gastrointestinal bypass surgery. *Proceedings of the National Academy of Sciences* (impact factor: 9.68). 05/2013; DOI:10.1073/pnas.1216959110 Source: PubMed.

95. *Ibid*. 92.

96. Pembrey M., Bygren, L.,, Kaati, G., Edvinsson, S., Northstone, K., Sjöström, M., Golding, J.& ALSPAC Study Team (2006) Sex-specific, male-line transgenerational responses in humans. *Eur J Hum Genet*. Feb;14(2):159–66.

Chapter 8

97. Zak, Paul J. (2012) *The Moral Molecule. The Source of Love and Prosperity*. Dutton, Published by Penguin Group (USA) Inc. New York, New York

98. Zak, P.J., Kurzhan, R., & Matzner, W. T. (2005). Oxytocin is associated with human trustworthiness. *Hormones and Behavior* 48, 522–27.

99. The synapse is the space between two neighboring cells. These proteins are found on the receiving side of this space.

100. Richards, G., Simionato, E., Perron, M., Adamska, M., Vervoort, M. & Degnan, B. (2008) Sponge genes provide new insight into the evolutionary origin of the neurogenic circuit. *Curr. Biol. 18, 1156–1161*

101. Kass-Simon, G. & Pierobon, P. (2007) Cnidarian chemical neurotransmission, an updated overview. Comparative Biochemistry and Physiology—Part A: *Molecular & Integrative Physiology*, Volume 146, Issue 1, Pages 9–25.

102. Ribeiro, P., El-Shehabi, F. & Patocka, N. (2005) Classical transmitters and their receptors in flatworms. *Parasitology* 131, S19-S40 doi:10.1017/ S0031182005008565.

103. Myers, David G. *Psychology 4th Edition.*New York:Worth Publishers Inc,1995: 43.

104. There are actually hundreds of different sense receptors besides the five classics we are taught in grade school, but for the sake of brevity and common knowledge I will just list the five.

105. MacLennon, Bruce J. (2011) Protophenomna and their Physical Correlates. *Journal of Cosmology*, 2011, Vol 14.

106. Zak, Paul J. (2012) *The Moral Molecule. The Source of Love and Prosperity.* Dutton, Published by Penguin Group (USA) Inc. New York, New York.

Chapter 9

107. Mlodinow, L. *Subliminal: How Your Unconscious Mind Rules Your Behavior*. Vintage Books. A division of Random House, Inc. New York. 2012

108. Nowakowski, R.S. (2006) Stable neuron numbers from cradle to grave. *Proceedings of the National Academy of Sciences of the United States of America.* 103(33):12219–12220.

109. Kandel, E.R. (2001) The Molecular Biology of Memory Storage: A Dialogue Between Genes and Synapses. *Science* Vol 294.

110. *Ibid.* 108.

111. Scripps Research Institute (2007, September 6). Specific Neurons Involved In Memory Formation Identified. *ScienceDaily*

112. Atkinson, R. C., & Shiffrin, R. M. (1971). The control processes of short-term memory. *Institute for Mathematical Studies in the Social Sciences,* Stanford University.

113. Miller, G. (1956). The magical number seven, plus or minus two: Some limits on our capacity for processing information. *The psychological review,* 63, 81–97.

114. Cowan, N. (2000). The magical number 4 in short-term memory: A reconsideration of mental storage capacity. *Behavioral and Brain Sciences,* 24, 87–185

115. Darnell, A. (2012) Short Term Memory: Age Related Changes During Childhood. *Voices*

116. Baddeley A. (2003). "Working memory: looking back and looking forward". *Nature Reviews. Neuroscience* 4 (10): 829–39

117. McLeod, S. A. (2008). *Working Memory—Simply Psychology.* Retrieved from http://www.simplypsychology.org/working%20memory.h tml

118. Tarnow, Eugen (2008) Short Term Memory May Be the Depletion of the Readily Releasable Pool of Presynaptic Neurotransmitter Vesicles. http://cogprints.org/6317/

119. Memory, short term. WikEd. 4/2012

120. *Ibid.* 108, 1031–1032.

121. *Ibid.* 108, 1032.

122. Kelleher, R.J., Govindarajan, A., Jung, H.-Y., Kang, H., & Tonegawa, S. (2004) Translational control by MAPK signaling in long-term synaptic plasticity and memory. *Cell* 116:467–479.

123. Thomson, E. (2004) Team discovers memory formation mechanism. *News Office*. February 11.

124. Scharf, M., Woo, N., Lattal, K., Young, J., Nguyen, P. & Abel T. (2002) Protein synthesis is required for the enhancement of long-term potentiation and long-term memory by spaced training. *J Neurophysiol.* Jun;87(6):2770–7.

125. *Ibid.* 108, 1034.

126. Martin, K., Casadio, A., Zhu, H., E, Y., Rose, J. Chen, M., Baily, C.& Kandel, E. (1997) Synapse-Specific, Long-Term Facilitation of Aplysia Sensory to Motor Synapses: A Function for Local Protein Synthesis in Memory Storage. *Cell* Vol. 91, issue 7, pages 927–938

127. *Ibid.* 122.

128. *Ibid.* 108.

129. David-Benjamin, G., Akatal, D & Davis, R. (2011) The Long-Term Memory Trace Formed in the Drosphila a/b Mushroom Body Neurons is Abolished in Long-Term Memory Mutants. *The Journal of Nueroscience,* 31(15):5643–5647.

130. Lynch, M. (2004) Long-Term Potentiation and Memory. *Physiological Reviews,* vol. 84 no. 1 87–136.

131. *Ibid.* 129.

132. Mechanism of Long-Term Memory Identified. (2011) *ScienceDaily,* April 13.

133. *Ibid.*

134. Ronald Davis quoted from the article in endnote 25.

135. *Ibid.* 108, 1037.

136. O'Keefe, J. & Dostrovsky, J. (1971) The hippocampus as a spatial map. Preliminary evidence from unit activity in the freely-moving rat. *Brain Res.* Nov;34(1):171–5.

137. Ekstrom, A., Kahana, M., Caplan, J., Fields, T., Isham, E., Newman, E. & Fried, I. (2003). "Cellular networks underlying human spatial navigation". *Nature* 425 (6954): 184–8.

138. Hassibis, D., Chu, C., Rees, G., Weiskopf, N., Molyneux, P. &, Maguire, E.. (2009) Decoding neuronal ensembles in the human hippocampus. *Current Biology,* 12 March.

139. 'Mind-Reading' Experiment Highlights How Brain Records Memories. (2009) *ScienceDaily,* March 12,.

140. Scripps Research Institute (2007, September 6). Specific Neurons Involved In Memory Formation Identified. *ScienceDaily.*

141. Seung, S. (2012) Connectome: How the Brain's Wiring Makes Us Who We Are. Houghton Mifflin Harcourt Publishing Company. New York, New York. 142. This is a completely fictional account of what might occur if you were sitting there watching Jeopardy.

143. McPherson, F. The Role of Emotion in Memory. Mempower. http://www.memory-key.com/memory/emotion

144. Canli,T., Desmon, J., Zhao, Z. & Gabriela, J. (2002) Sex differences in the neural basis of emotional memories. *Proceedings of the National Academy of Sciences*, 99, 10789–10794.

145. Charles, S., Mather, M. & Cartensen, L. (2003) Aging and Emotional Memeory: The Forgettable Nature of Negative Images for Older Adults. *Journal of Experimental Psychology: General, 132*(2), 310–24.

146. The neocortex is the outer most layer of the brain. It is a six-layered structure thought to be the source of our consciousness.

147. Buzsaki, G. (2010). Neural syntax: assembly sequences, synapsembles and readers. *Neuron*

148. McGaugh, J. (2000) Panel: The Science of Memory and Emotion "How Emotions Strenghthen Memory" *Library of Congress.*

149. Kauppi, K., Nilsson, L., Adolfsson, R., Erickson, E., & Nyberg, L. (2011) KIBRA Polymorphism Is Related to Enhanced memory and Elevated Hippocampal Processing. *Journal of Neuroscience,* 31 (40):14218 doi: 10.1523/jneurosci.3292–11.2011.

Chapter 10

150. Joseph, R. (2000) Olfactory Limbic System, From: *Neuropsychiatry, Neuropsychology, Clinical Neuroscience* (3rd Ed). Academic Press, New York.

151. Joseph, R. (2000) Amygdala, From: *Neuropsychiatry, Neuropsychology, Clinical Neuroscience* (3rd Ed).

Academic Press, New York.

152. Dreifuss, J., Murphy, J., & Gloor, P. (1968) Contrasting effects of two identified amygdaloid efferent pathways on single hypothalamic neurons. *Journal of Neurophysiology*, 31, 237.

153. Zak, P. (2012) *The Moral Molecule: The Source of Love and Prosperity.* Published by Dutton, a member of Penguin Group USA. New York, New York.

154. Crusco, A. & Wetzel, C. (1984). The Midas Touch: The effects of interpersonal touch on restaurant tipping. *Personality and Social Psychology Bulletin*, 10, 4, 512–517.

155. Guergen, N. (2007) Courtship Compliance: The effect of touch on women's behavior. *Social Influence,* volume 2, Issue 2.

156. Williams, L. & Bargh, J. (2008). Experiencing physical warmth promotes interpersonal warmth. *Science,* 322(5901), 606–607. doi:10.1126/science.1162548

157. Ackerman, J., Nocera, C., Bargh, J. (2010). Incidental Haptic Sensations Influence Social Judgments and Decisions. *Science* 25 Vol. 328 no. 5986 pp. 1712–1715 doi: 10.1126/science.1189993

158. Amaral D., Price, J., Pitkanen, A., & Carmichael, S. (1992) "Anatomical organization of the primate amygdala complex." In: *The Amygdala,* J. Aggleton, ed., New York: Wiley-Liss, pp 1–67.

159. *Ibid.* 151.

160. Fallon J. & Ciofi P. (1992). "Distribution of monoamines within the amygdala," in *The Amygdala: Neurobiological Aspects of Emotion, Memory, and Mental Dysfunction,* ed. Aggleton J. P., editor. (New York: Wiley-Liss, Inc;), 97–114.

161. LeDoux, J. (2008) Amygdala. *Scholarpedia*

162. Hamann, S. (2005). Sex Differences in the Responses of the Human Amygdala. *The Neuroscientist,* Sage Publications.

163. *Ibid.* 151.

164. *Idid.* 151.

165. Cahill, L., Haier, R. Fallon, J., Kilpatrick, L., Lawrence, C., Potkin, S. & Alkire, M.. (2001) Sex-related differences in amygdala activity during emotionally influenced memory storage. *Neurobiology of Learning and Memory* 75:1 9.

166. Seidhtz, L. & Deiner, E. (1998) Sex differences in the recall of affective experiences. *J Pers Soc Psychol* 74:262 71.

167. Karama, S., Lecours, A., Leroux, J., Bourgouin, P., Beaudoin, G., Joubert,S., & Beauregard, M. (2002) Areas of brain activation in males and females during viewing of erotic film excerpts. Human Brain Map 16:1 13

168. *Ibid.* 151.

169. Holstege, G., Georgiadis, J., Parans, A., Meiners, L., an der Graaf, H., & Reinders, A.. (2003) Brain activation during human male ejaculation. *J Neuroscience* 23:9185 93.

170. Symons D. (1979) *The evolution of human sexuality.* Oxford, UK: Oxford University Press.

171. Swaab, D. (2008) Sexual Orientation and its basis in brain structure and function. *Proceeding of the National Academy of Science.* Vol. 105, no. 30 10273–10274.

172. Puts, D., Jordan, C., & Breedlove, S. (2006) O brother, where art thou? The fraternal birth-order effect on male sexual orientation. *Proceedings of the National Academy of Science.* vol. 103 no.28 10531–10532.

Chapter 11

173. Gupta, R., Koscik, T., Bechera, A., & Tranel, D. (2011) The amygdala and decision making. *Neuropsycologia.* 49(4): 760–6.

174. Adolps, R, & Tanel, D. (2003) Amygdala damage impairs recognition from scenes only when they contain facial expressions. *Neuropsychologia* 41(10):1281–9.

175. LeDoux, J. (1996) *The Emotional Brain.* Simon and Schuster, New York, New York.

176. Bechara, A., Damasio, H., Tranel, D., & Damasio, A. (1997). "Deciding advantageously before knowing the advantageous strategy". *Science* 275 (5304): 1293–1294. doi:10.1126/science.275.5304.1293

177. Gladwell, M (2005) *Blink: The Power of Thinking without Thinking.* Back Bay Books, New York, New York. A fascinating read with many scenarios of how the first inclination is usually the right one.

178. Ambady, N, & Rosenthal, R. (1993) Half a Minute: Predicting Teacher Evaluations From Thin Slices of Nonverbal Behavior and Physical Attractiveness. *Journal of Personality and Social Psychology,* Vol. 64, No. 3, 431–441.

179. Damasio, A. (1994) *Descartes' Error: Emotion, Reason, and the Human Brain,* Putnam

180. Joseph, R. (2000) Charles Whitman: The Amygdala and Mass Murder. brainmind.com

181. Charles J. Whitman Catastrophe, Medical Aspects. Report to Governor, 9/8/66.

182. O'keefe, J. & Bouma, H., (1969) Complex sensory properties of certain amygdala units in freely moving cats. *Experimental Neurology*, Vol. 23, no. 3, 384–398.

183. Joseph, R. (2000) Amygdala, From: *Neuropsychiatry, Neuropsychology, Clinical Neuroscience* 3rd Edition. Academic Press, New York.

184. Joseph, R. (1999) Frontal lobe psychopathology: Mania, depression, aphasia, confabulation, catatonia, preservation, obsessive compulsive, schizophrenia. *Journal of Psychiatry*, 62, 138–172

Chapter 12

185. Zak, P., Stanton, A. & Ahmadi, S. (2007). Oxytocin increases generosity in humans. *Public Library of Science One* 2(11), e1128. Doi:10.1371/journal.pone.0001128.

186. Edelman, G. (1992) Linking Brain to Behavior: Value and selection in neural populations. In *Proceedings of the Course on Neuropsychology: The Neuronal Basis of Cognitive Function*, Fidia Research Foundation, pp. 55–66, Thieme Medical Publishers, Inc., New York.

187. Pink Floyd, The Wall (1982), *Mother.*

188. Seung, S. (2012) *Connectome: How the Brain's Wiring Makes Us Who We Are.* Houghton Mifflin Harcourt Publishing Company. New York, New York.

Chapter 13

189. Laplace, P., (1902) *A Philosophical Essay*, New York

190. Einstein, A. (1946) *Autobiographical notes.* In Albert Einstein, Alice Calaprice, Freeman Dyson , *The Ultimate Quotable Einstein* (2011), 397

191. Hawking, S. (1988). A Brief history of Time: From the Big Bang to Black Holes. Bantam Books.

192. Joseph, R. (2011) Origins of thought: Language, Egocentric Speech and the Multiplicity of Mind. *Journal of Cosmology*, Vol. 14.

193. Krause, L. (2012) *The Universe from Nothing.* Simon and Schuster LTD. London

194. A description of this study can be found in almost any Microbiological text. By putting a bend into a narrow tube Pasteur was able to show bacteria from the air was responsible for media contamination, not that cells just came into existence.

Chapter 14

195. Joseph, R. (2011) Origins of thought: Language, Egocentric Speech and the Multiplicity of Mind. *Journal of Cosmology*, Vol. 14.

196. *Ibid.*

197. *Ibid.*

198. Woodward, A. (2003) Infants' developing understanding of the link between looker and object. *Dev Sci*,;6(3):297–311.

199. Society of Neuroscience (2008) Brain Briefings—Mirror Neurons. Society of Neuroscience, Washington DC 20005.

200. Iacobani M, Molnar-Szakacs, I., Gallese, V., Buccini, G., Mazziotta, J., Rizzolatti, G. (2005) Grasping the intention of others with one's own mirror neuron system. *PLos Biology*, 3(3):e79

201. Enticott, P., Johnston, P., Herring, S., Hoy, K., & Fitzgerald, P. (2008) Mirror Neuron activation is associated with facial emotion processing. *Neuropsychologia*; 46(11):2851–2854.

Chapter 15

202. Frackowiak, R. (2004), *Functional brain imaging* (2nd ed.), London: Elsevier Science.

203. Reiss, D. & Marino, L. (2001) Mirror self-recognition in the bottle-nosed dolphin: A case of cognitive convergence. Article—*Proceedings of the National Acadamy of Sciences of the United States of America*, 98(10), 5937–5942.

204. Hyatt, C. & Hopkens, W. (1994) Self-awareness in bonobos and chimpanzees: a comparative approach. In *Self-awareness in animals and humans: developmental perspectives*, S.T. Parker and R.W. Mitchell and M.L. Boccia (eds) pp 248–253. New York: Cambridge University Press.

205. Prior, H., Schwarz, A. & Güntürkün, O. (2008) Mirror induced behavior in magpie (pica pica): Evidence of Self-recognition. *Plos Biology,* 6, 1642–1650.

206. Gallup, G. (1970) Chimpanzees: Self recognition. *Science,* 167, 86–87.

207. Nani, A., Eddy, C. & Cavanna, A. (2011) The Quest for Animal Consciousness. *Journal of Cosmology,* vol.14.

208. Lipton, B. (2005) *The Biology of Belief: Unleashing the Power of Consciousness, Matter & Miracles.* Hay House, Inc. Carlsbad, California.

Chapter 16

209. Children do not have sufficient neuronal connections between hemispheres to be able to fully consider consequences for their actions.

210. Bugliosi, V. (2008) *Outrage: The Five Reasons OJ Simpson Got Away with Murder,* W.W. Norton & Company. 500 Fifth Ave. New York, NY.

211. Zak, P. (2012) *The Moral Molecule: The Source of Love and Prosperity.* Published by Dutton, a member of Penguin Group USA. New York, New York.

212. Gasquoine, P. (1993). Bilateral alien hand signs following destruction of the medial frontal cortices. *Neuropsychiatry, Neuropsychology & Behavioral Neurology,* 6, 49–53.

213. Baron-Cohen, S. Bolton, P., Wheelwright, S., Scahill, V., Short, L., Mead, G. & Smith, A. (1998). Autism occurs more often in families of physicists, engineers, and mathematicians. *Autism* 2,296–301.

214. Joseph, R. (2000) Amygdala, From: *Neuropsychiatry, Neuropsychology, Clinical Neuroscience* 3rd Edition. Academic Press, New York.

215. Complied from website: www.calgaryautism.com/characteristics.htm

216. Baron-Cohen, S, (2006) Two new theories of autism: hyper-systemizing and assortative mating. *Archives of Diseases in Childhood,* 91, 2–5.

217. Baron-Cohen, S., Wheelwright, S., Spong, A., Scahill, V. & Lawson, J. The Autism-Spectrum Quotient (AQ): Evidence from Asperger Syndrome/High-Functioning Autism, Males and Females, Scientists and Mathematicians *J. Dev. Learn. Disord.* 5, 47–78 (2001).

218. *Ibid.* 213.

219. Braunschweig, D., Krakowiak, P., Duncanson, P., Boyce, R., Hansen, R., Ashwood, P., Hertz-Piciotto, I., Pessah, I. & Van der Water, J. (2013) Autism-specific maternal antibodies recognize critical proteins in developing brain. *Translational Psychiatry*, 3:e277.

220. NBC sitcom *The Big bang Theory*.

Chapter 17

221. Mlodinow, L. (2012). *Subliminal: How Your Unconscious Mind Rules Your Behavior*. Vintage Books. New York.

222. Lipton, B. (2008). *The Biology of Belief. Unleashing the Power of Consciousness, Matter & Miracles*. Hay House Inc. Carlsbad, California.

223. Seung, S. (2012) *Connectome: How the Brain's Wiring Makes Us Who We Are*. Houghton Mifflin Harcourt Publishing Company. New York, New York.

224. Fried, I.,Mukamel, R., & Kreiman, G. (2010) Internally Generated Preactivation of Single Neurons in Human Medial Frontal Cortex Predicts Volition. *Neuron* 69, 548–562.

225. Joseph, R. (2011). The Neuroanatomy of Free Will, Loss of Will, Against the Will, Alien Hand" *Journal of Cosmology*, Vol 14. Included as a collection of papers taken from the Journal of Cosmology titled Consciousness and the Universe: Quantum Mechanics, Evolution, Brain & Mind. Cosmology Science Publishers, Cambridge MA

226. *Ibid*.

227. Bauby, J. (1998). *The Diving Bell and the Butterfly: A memoir of Life in Death*. Vintage.

228. Lhermite, F. (1983). "Utilization Behavior" and its relation to lesions of the frontal lobes. *Brain*, 106, 237–255

229. *Ibid*. 225.

230. Gasquoine P. (1993) Alien hand sign. *J Clin Exp Neuropsychol*, 15:653–67.

231. Goldberg, G., Mayer, H., & Toglia, J. (1981) Medial frontal cortex infarction and the alien hand sign. *Arch Neurol*, 38:683–6.

232. *Ibid*. 225.

233. Laibow, R. (1999). "Medical applications of neurofeedback" in *Introduction to Quantitative EEG and Neurofeedback.* J. R. Evans & A. Abarbanel,(eds.) Academic Press, New York.

234. Bounias, M. Laibow, R., Bonaly, A. & Stubblebine, A. (2002). Q-EEG Neurofeedback of Brain Injured Patients, Part I: Typological Classification of Clinical Syndromes, *J. NeuroTherapy*

235. Joseph, R. (2011). The Split Brain: Two Brains—Two Minds. In Lana Tao (Ed.), *Consciousness and the Universe, Quantum Physics, Evolution, Brain and Mind. Journal of Cosmology,* Cosmology Science Publishers, Cambridge, MA.

236. Yakolvlev, P. L., & Lecours, A. (1967). The myelogenetic cycles of regional maturation of the brain. In A. Minkowski (ed.), Regional development of the brain in early life, (pp.404–491). London Blackwell

237. *Ibid.* 234, pp131–137.

Chapter 18

238. Piazza J., Bering, J. & Ingram, G. (2011). "Princess Alice is watching you": Children's belief in an invisible person inhibits cheating. *Journal of Experimental Child Psychology,* doi:10.1016/J.Jeep.2011.02.003.

239. Bering, J. & Parker, B. (2006). Children's attributions of intentions to an invisible agent. *Developmental Psychology,* 42, 253–262

240. Bering, J. (2005). The evolutionary history of an illusion: Religious causal beliefs in children and adults. In B. Ellis & D. Bjorklund (Eds.), *Origins of the Social Mind: Evolutionary Psychology and Child Development* (pp. 411–437). New York: Guilford Press.

241. *Ibid.* 238.

242. Gould, S. & Lewontin, R. (1979) "The Spandrels of San Marco and the Panglossian Paradigm: A Critique of the Adaptationist Programme" *Proc. Roy. Soc. London B* 205 pp. 581–598.

INDEX

T

U

V

W

Y

Z

CPSIA information can be obtained
at www.ICGtesting.com
Printed in the USA
FSOW02n0720300118
43970FS